Writings on an
Ethical Life

Also by Peter Singer

Democracy and Disobedience

Animal Liberation

Practical Ethics

Marx

Animal Factories
(with Jim Mason)

The Expanding Circle

Hegel

The Reproduction Revolution
(with Deane Wells)

Should the Baby Live?
(with Helga Kuhse)

Animal Liberation: A Graphic Guide
(with Lori Gruen)

Save the Animals!
(with Barbara Dover and Ingrid Newkirk)

How Are We to Live

The Great Ape Project
(editor with Paola Cavalieri)

Rethinking Life and Death

Ethics into Action

A Darwinian Left

An *Imprint of* HarperCollins*Publishers*

Writings on an Ethical Life

PETER SINGER

A hardcover edition of this book was published in 2000 by The Ecco Press.

HarperCollins books may be purchased for educational, business, or sales promotional use. For information please write: Special Markets Department, HarperCollins Publishers Inc., 10 East 53rd Street, New York, NY 10022.

First Ecco paperback edition published in September 2001

Designed by Cassandra J. Pappas

The Library of Congress has catalogued the hardcover edition as follows:
Singer, Peter, 1946–
[Selections. 2000]
Writings on an ethical life / Peter Singer.—1st ed.
p. cm.
Includes bibliographical references.
ISBN 0-06-019838-9
1. Ethics. I. Title.
BJ354.S561 2000
170—dc21 00–042674

ISBN 0-06-000744-3 (pbk.)

02 03 04 05 DC/RRD 10 9 8 7 6 5 4

Contents

Acknowledgments

"Moral Experts" first appeared in *Analysis,* vol. 32 (1972), pp. 115–117. "About Ethics" from *Practical Ethics* (Cambridge University Press, Cambridge, 2nd ed., 1993), pp. 1–14. "Preface to the 1975 Edition" from *Animal Liberation* (New York Review/Random House, New York, 1975), pp. vii–xiv. "All Animals Are Equal . . ." from *Animal Liberation* (New York Review/Random House, New York, 2nd ed., 1990), pp. 1–21. "Tools for Research" from *Animal Liberation,* 2nd ed., pp. 40–42, 45–47, 81–83, 85–86, 90–92, 94. "Down on the Factory Farm . . ." from *Animal Liberation,* 2nd ed., pp. 95–97, 129–136. "A Vegetarian Philosophy" from *Consuming Passions: Food in the Age of Anxiety* (Manchester University Press, Manchester, 1998), pp. 72–75, 77–80. Passages from "Bridging the Gap" have appeared in *Rethinking Life and Death: The Collapse of Our Traditional Ethics* (Text Publishing, Melbourne, 1994; St. Martin's Press, New York, 1995) and in "The Rights of Ape," *BBC Wildlife,* June 1993, pp. 28–32. "Environmental Values" from *The Environmental Challenge* (Lanman Cheshire, Melbourne, 1991), pp. 3–24. "Famine, Affluence, and Morality," from *Philosophy and Public Affairs* (Spring 1972), vol. 1, pp. 229–243. "The Singer Solution to World Poverty" from *The New York Times Magazine,* September 5, 1999. "What's Wrong with Killing?" from *Practical Ethics,* 2nd ed., pp. 83–109. "Taking Life: The Embryo and the Fetus" from *Practical Ethics,* 2nd ed., pp. 138–174. "Prologue" from *Rethinking Life and Death,* pp. 1–6. "Is the Sanctity of Life Ethic Terminally Ill?" from *Bioethics* vol. 9, no. 3, p. 4 (1995), pp. 327–343 and, in part, in *Rethinking Life and Death.* "Justifying Infanticide" from *Practical Ethics,* 2nd ed., pp. 181–191. "Justifying Voluntary Euthanasia" from *Practical Ethics,* 2nd ed., pp. 193–200. "Euthanasia: Emerging from

Hitler's Shadow" was delivered at the Third World Congress of the International Association of Bioethics, San Francisco, 1996. "In Place of the Old Ethic" from *Rethinking Life and Death*, pp. 187–222. "The Ultimate Choice" from *How Are We to Live?* (Text Publishing, Melbourne, 1993; Prometheus: Amherst, NY, 1995), pp. 1–20. "Living Ethically" from *How Are We to Live?* pp. 154–170. "The Good Life" from *How Are We to Live?* pp. 225–235. "Darwin for the Left" from *Prospect*, June 1998, pp. 26–30; portions of this essay also appear in A *Darwinian Left* (Weidenfeld and Nicolson, London, 1990; Yale University Press, New Haven, CT, 2000). "A Meaningful Life" from *Ethics into Action: Henry Spira and the Animal Rights Movement* (Rowman & Littlefield, Lanham, MD, 1998), chapter 6. "Animal Liberation: A Personal View" from *Between the Species*, vol. 2 (Summer 1986), pp. 148–154. "On Being Silenced in Germany" from *The New York Review of Books*, August 15, 1991. "An Interview" is excerpted from an interview aired on September 10, 1999, on WNET-TV and an interview that appeared in *New Scientist*, January 8, 2000.

Introduction

HAD THERE BEEN no protest rallies at the gates of Princeton University, had the superrich aspiring presidential candidate Steve Forbes not vowed to freeze his donations to his alma mater until it got rid of me, had my appointment as DeCamp Professor of Bioethics at Princeton not turned into what *The New York Times* called the biggest academic commotion since an American university tried to hire the notorious advocate of free love, Bertrand Russell, the book you now have in your hands would not have existed. For without those events the media would barely have noticed my arrival in the United States, my controversial views on a variety of ethical issues would not have become the topic of conversation around the country, and it could never have occurred to Daniel Halpern at The Ecco Press that while everyone was discussing Peter Singer's views, the discussions were based largely on short quotations and secondhand summaries. Many had strong opinions about my work but few had actually read any of my books or articles. They did, however, have the excuse that my writings are scattered across a number of books and journals, not all of them easy to obtain. What people needed, Dan thought, was a handy, one-volume selection that would present my central ideas, in my own words and with sufficient context to enable them to be understood. So here it is.

What are the ideas that have caused so much controversy? They are to be found in the pages of this book, and I would prefer them to be read in their context, not in bald summary. I have spent close to thirty years working in practical ethics, which means that when I began there was no such field. The study of ethics, in philosophy departments in the English-speaking world, was then focused on the analysis of moral language and was supposed to be morally neutral; that is, it did not lead to any judgments about anything's being right or wrong, good or bad. A moral philosopher, so the standard view went, is not in any way an expert on moral issues. The short essay entitled "Moral Experts" that immediately follows this introduction was written to challenge this conception of the subject matter of ethics. Since its appearance in 1972, the field has changed dramatically, though not, I am fairly sure, because of any influence that essay may have had. More probably, the article was a sign of times that were already changing. The momentous issues of civil rights, racial equality, and opposition to the war in Vietnam had made the subject matter of most university ethics courses seem dry and insignificant and to spend time studying them, self-indulgent. There were more important things to do. Students were demanding "relevance," and philosophers began to realize that their area of expertise did, after all, have something to say on how we can find answers to such fundamental and perennially important questions as why racial discrimination is wrong, whether we are under an obligation to obey an unjust law, and what, if anything, can make it right to go to war.

The issue on which I have made my most significant contribution to ethical thinking took a little longer to become prominent. Claims that I am "the most influential living philosopher" generally refer to my work on the ethics of our relations with animals. There is an element of media hype in such claims, of course, but the grain of truth in them is the fact that my book *Animal Liberation* played a significant role in kicking off the modern animal rights movement, and very few living philosophers have had their ideas taken up in that way. In contrast, my views on the obligation of the rich to help the world's poorest people are potentially as significant as my thinking about animals, but they have, unfortunately, had much less influence. The issues raised by the critical work I have done on the idea of the sanctity of human life, including the discussion of euthanasia that has aroused such hostility, are in one respect less important than those two topics, simply because the treatment of animals and maldistribution of wealth affect far more people (or in the case of the animals, sentient beings), and relatively simple changes in those areas could relieve so much suffering. It is my critique of the sanctity of human life that gets the media headlines, however, because it can easily be made to sound quite shocking and there is no problem in finding people who will strenuously oppose it. Hence it provides the controversy on which the media feed.

All of these views have a common core. They rest on four quite simple claims:

1. Pain is bad, and similar amounts of pain are equally bad, no matter whose pain it might be. By "pain" here I would include suffering and distress of all kinds. This does not mean that pain is the only thing that is bad, or that inflicting pain is always wrong. Sometimes it may be necessary to inflict pain and suffering on oneself or others. We do this to ourselves when we go to the dentist, and we do it to others when we reprimand a child or jail a criminal. But this is justified because it will lead to less suffering in the long run; the pain is still in itself a bad thing. Conversely, pleasure and happiness are good, no matter whose pleasure or happiness they might be, although doing things in order to gain pleasure or happiness may be wrong, for example, if doing so harms others.

2. Humans are not the only beings capable of feeling pain or of suffering. Most nonhuman animals—certainly all the mammals and birds that we habitually eat, like cows, pigs, sheep, and chickens—can feel pain. Many of them can also experience other forms of suffering, for instance, the distress that a mother feels when separated from her child, or the boredom that comes from being locked up in a cage with nothing to do all day except eat and sleep. Of course, the nature of the beings will affect how much pain they suffer in any given situation.

3. When we consider how serious it is to take a life, we should look, not at the race, sex, or species to which that being belongs, but at the characteristics of the individual being killed, for example, its own desires about continuing to live, or the kind of life it is capable of leading.

4. We are responsible not only for what we do but also for what we could have prevented. We would never kill a stranger, but we may

> know that our intervention will save the lives of many strangers in a distant country, and yet do nothing. We do not then think ourselves in any way responsible for the deaths of these strangers. This is a mistake. We should consider the consequences both of what we do and of what we decide not to do.

To most people these claims are not, in themselves, shocking. In some respects they seem like common sense. But consider the conclusions to which they point. Put together the first and the last, and add in some facts about the suffering caused by extreme poverty in the world's least developed countries, and about our ability to reduce that suffering by donating money to organizations that assist people to lift themselves out of this poverty. Consider, for example, the fact that the sum that buys us a meal in a fine restaurant would be enough to provide basic health care to several children who might otherwise die of easily preventable diseases. It follows from the first of my claims that the suffering of these children, or their parents, is as bad as our own suffering would be in similar circumstances, and it follows from the last claim that we cannot escape responsibility for this suffering by the fact that we have done nothing to bring it about. Where so many are in such great need, indulgence in luxury is not morally neutral, and the fact that we have not killed anyone is not enough to make us morally decent citizens of the world.

From the first and second premises I draw the conclusion that we have no right to discount the interests of nonhuman animals simply because, for example, we like the taste of their flesh. Modern industrialized agriculture treats animals as if they were things, putting them indoors and confining them whenever it turns out to be cheaper to do so, with no regard at all paid to their suffering or distress, as long as it does not mean that they cease to be productive. But we cannot ethically disregard the interests of other beings merely because they are not members of our species. Note that this argument says nothing at all about whether it is wrong to kill nonhuman animals for food. It is based entirely on the suffering that we inflict on farm animals when we raise them by the methods that are standard today.

The third premise is probably the most controversial because we are so used to thinking of the killing of a member of our own species as invariably much more serious than the killing of a member of any other species. But why should that be? Mere difference of species is surely not a morally significant difference. Suppose that there were Martians, just like us in respect of their abilities to think and to care for others, in their sense of justice, and in any other capacities we care to name, but of course, not members of the species *Homo sapiens*. We surely could not claim that it was all right to kill them, simply because of the difference of species. We might, of course, try to find some other differences, of greater moral relevance, between humans and members of other species. If, for example, we think that it is more serious to kill a human being than it is to kill a nonhuman animal, and we hold this view because we believe that every human being—and no other earthly creature—has an immortal soul, then our position is not contrary to the third premise, for it is taking the view that there is a characteristic—possession of an immortal soul—that makes it worse to kill some beings than others, and that happens to be possessed by all and only members of our species. My disagreement with that position is simply that I see no evidence for belief in an immortal soul, let alone one that happens to be the exclusive property of our species.

The third premise helps to explain what is true and what is misleading in the common assertion that I think the life of a human being is of no greater value than the life of an animal. It is true that I do not think that the fact that the human is a member of the species *Homo sapiens* is *in itself* a reason for regarding his or her life as being of greater value than that of a member of a different species. But, as I argue in more detail, in the extracts from *Practical Ethics* and *Rethinking Life and Death*, human beings typically, though not invariably, do have desires about going on living that nonhuman animals are not capable of having, and that does make a difference. Thus, I have no doubt that the events that we read about all too often in our newspapers, when someone gets a gun and starts randomly killing people in a school, church, or supermarket, are more tragic than the shooting of a number of animals in a field would be.

Though the third premise must be part of any full grounding of my views about why it is worse to kill some beings than others, some of my claims about euthanasia could be derived from the first and fourth premises alone, combined with widely accepted medical practice. Many doctors and theologians, including those who are quite conservative in their moral thinking, agree that when a patient's prospects of a minimally decent quality of life are very poor, and there is no likelihood of improvement, we are not obliged to do everything we could to prolong life. For example, if a baby is born with severe disabilities incompatible with an acceptable quality of life, and the baby then develops an infection, many doctors and theologians would say that it is permissible to refrain from giving the baby antibiotics. But these same people think that it would be wrong to allow the doctor to give the baby a lethal injection. Why? The motives, the intention, and the outcome may be the same in both cases. If it is sometimes right deliberately to allow a baby to die when a simple medical intervention could save its life, then it must also sometimes be right to kill the baby. To deny this is to refuse to take responsibility for deciding not to act, even when the consequences of omission and action are the same.

That this final conclusion is contrary to widely shared moral views, I readily concede. But the aim of practical ethics is not to produce a theory that will match all of our conventional moral responses, and thus confirm us in the views we already hold. Those responses come from many different sources. Sometimes they vary according to the customs of the society that we come from. Even when they are more or less universal among human societies, they may be no more than a reflection of the interests of the dominant group. That was true of justifications of slavery that for centuries were predominant in European slave-holding societies. It remains true, in many parts of the world, of the view that a married woman should obey her husband. It is not too big a step to see the same self-interested factors at work in our common moral views about the ways in which we may use animals. Here, as on many other moral issues, Christianity has for two thousand years been a powerful influence on the moral intuitions of people in Western societies. People do not need to continue to hold religious beliefs to be under the influence of Christian moral teaching. Yet without the religious beliefs—for example, that God created the world, that he gave us dominion over the other animals, that we alone of all of his creation have an immortal soul—the moral teachings just hang in the air, without foundations. If no better foundations can be provided for these teachings, we need to consider alternative views. So it is with the question of euthanasia, which, along with suicide, has in non-Christian societies like ancient Rome and Japan been considered both a reasonable and an honorable way of ending one's life. The shock with which some people react to any suggestion of euthanasia should therefore be not the end of the argument but a spur to reflection and critical scrutiny.

In choosing the passages that follow I have tried to convey what is essential to my thinking, leaving out a number of other works—for example, my first book, on whether there is an obligation to obey the law in a democracy; my short introductions to Hegel and Marx; and my writings on new reproductive technology. The omissions are not because I am no longer content with what I did in those books. On the contrary, I still think that, for example, being able to convey the gist of Hegel's philosophy to the general reader in less than one hundred pages is an achievement. But it does not mark me out from other skilled expositors of the great philosophers of the past. In the case of *Making Babies*, my coauthored book on the revolution in reproduction, the field has moved on since 1984, when it was first published, and the book needs to be rewritten rather than reprinted.

The fourth section of this book reflects an interest that I do think is essential to my work, though it has received less attention than my writings on animals, world hunger, or the sanctity of human life. The first substantial piece of work I did in philosophy, a thesis I wrote for the degree of master of arts, was on a question that is about ethics, rather than within ethics. Given that ethics can be very demanding, what are we to say to the amoralists, who ask why they should act ethically at all? I never felt that I had answered that question satisfactorily in my thesis, and I have returned to it on various occasions, but most fully in my book *How Are We to Live?* The question leads us to think about the ultimate values, the deepest goals, by which we live our lives, and here we tend to run up against the limits of philosophical argument. Is it still possible, at this fundamental level, to give reasons for choosing one way of life in preference to another? Is it all a matter of what will make us happier, or live a more meaningful and fulfilling life? Here we move across the ill-defined border between philosophy and psychology, and can no longer find chains of reasoning that should persuade any rational person. Were we incapable of empathy—of putting ourselves in the position of others and seeing that their suffering is like our own—then ethical reasoning would lead nowhere. If emotion without reason is blind, then reason without emotion is impotent.

Most of the extracts here come from works that I wrote with a general reader in mind, while trying to be sufficiently rigorous to be of interest to my philosophical colleagues. *Animal Liberation* has been by far my most successful book, in terms of reaching a wide audience, having sold close to half a million copies. It has also, in my admittedly partisan view, stood up well against extensive probing by philosophers, in journal articles too numerous to keep track of, and in several books, for example David DeGrazia's fine study of the field, *Taking Animals Seriously*. But the path that is trodden both by general readers and by academic philosophers is a narrow one, and I may have fallen off it from time to time, on one side or the other.

A similar tension exists in a more acute form when it comes to campaigning for a cause, while remaining true to the philosophical vocation. During the past twenty-five years there have been few times when I was not heading at least one organization struggling to improve the situation of animals, and I have also been involved with groups working to protect wilderness, aid some of the poorest people in the world, and legalize voluntary euthanasia, as well as standing for the Australian senate as a candidate of the Greens. From the point of view of a campaign, the complexities of a defensible ethical position can often be an obstacle. A friend in the animal movement once told me that I am not a street fighter. It is true that the most effective campaigners are often people of uncompromising temperament who take their stand on a principle to which they will allow no exceptions. They see only one side of any argument and regard their opponents as evil incarnate. The atmosphere of campaigning organizations is conducive to this kind of temperament. Take, for example, the use of animals in research. As the extracts from *Animal Liberation* to be found in this book illustrate, a great many experiments on animals cause them enormous suffering and do nothing more for humans than allow another detergent or food preservative to be marketed. But what if an experiment offered some hope of finding a cure for a major disease? For a group campaigning against animal experiments, everything becomes much simpler if we believe that no experiment on an animal can ever be justified, whatever the benefit to humans might be. Then there can be no awkward questions about the conditions under which an experiment might be justifiable. It is also easier to campaign against experimenters if one can convince oneself that they are all sadists who get their jollies from cutting up fully conscious kittens. Unfortunately, reality is not so black-and-white. Seeing the more complex picture leaves me liable to be attacked from both sides—by those who exploit animals, because I threaten what they do, and by the more hard-line members of the animal movement, for not taking the party line. The temptation is, then, to leave the campaigning to others and retreat to my desk. But I think it important, not just to write and teach but to try to make a difference in more immediate ways as well. That point is made in the extract in this collection from my recent book, *Ethics into Action*.

The Nature of Ethics

Moral Experts

FROM *Analysis*

THE FOLLOWING POSITION has been influential in recent moral philosophy: there is no such thing as moral expertise; in particular, moral philosophers are not moral experts. Leading philosophers have tended to say things like this:

> It is silly, as well as presumptuous, for any one type of philosopher to pose as the champion of virtue. And it is also one reason why many people find moral philosophy an unsatisfactory subject. For they mistakenly look to the moral philosopher for guidance.[1]

or like this:

> It is no part of the professional business of moral philosophers to tell people what they ought or ought not to do. . . . Moral philosophers, as such, have no special information not available to the general public, about what is right and what is wrong; nor have they any call to undertake those hortatory functions which are so adequately performed by clergymen, politicians, leader-writers . . .[2]

Assertions like these are common; arguments in support of them less so. The role of the moral philosopher is not the role of the preacher, we are told. But why not? The reason surely cannot be, as C. D. Broad seems to suggest, that the preacher is doing the job "so adequately." It is because those people who are regarded by the public as "moral leaders of the community" have done so badly that "morality," in the public mind, has come to mean a system of prohibitions against certain forms of sexual enjoyment.

Another possible reason for insisting that moral philosophers are not moral experts is the idea that moral judgments are purely emotive, and that reason has no part to play in their formation. Historically, this theory may have been important in shaping the conception of moral philosophy that we have today. Obviously, if anyone's moral views are as good as anyone else's, there can be no moral experts. Such a crude version of emotivism, however, is held by few philosophers now, if indeed it was ever widely held. Even the views of C. L. Stevenson do not imply that anyone's moral views are as good as anyone else's.

A more plausible argument against the possibility of moral expertise is to be found in Ryle's essay "On Forgetting the Difference between Right and Wrong," which appeared in *Essays in Moral Philosophy*, edited by A. Melden.[3] Ryle's point is that knowing the difference between right and wrong involves caring about it, so that it is not, in fact, really a case of knowing. One cannot, for instance, forget the difference between right and wrong. One can only cease to care about it. Therefore, according to Ryle, the honest man is not "even a bit of an expert at anything" (p. 157).

It is significant that Ryle says that "the honest man" is not an expert, and later he says the same of "the charitable man." His conclusion would have had less initial plausibility if he had said "the morally good man." Being honest and being charitable are often—though perhaps not as often as Ryle seems to think—comparatively simple matters, which we all can do, if we care about them. It is when, say, honesty clashes with charity (If a wealthy man overpays me, should I tell him, or give the money to famine relief?) that there is need for thought and argument. The morally good man must know how to resolve these conflicts of values. Caring about doing what is right is, of course, essential, but it is not enough, as the numerous historical examples of well-meaning but misguided men indicate.

Only if the moral code of one's society were perfect and undisputed, both in general principles and in their application to particular cases, would there be no need for the morally good man to be a thinking man. Then he could just live by the code, unreflectively. If, however, there is reason to believe that one's society does not have perfect norms, or if there are no agreed norms on a whole range of issues, the morally good man must try to think out for himself the question of what he ought to do. This "thinking out" is a difficult task. It requires, first, information. I may, for instance, be wondering whether it is right to eat meat. I would have a better chance of reaching the right decision, or at least a soundly based decision, if I knew a number of facts about the capacity of animals for suffering, and about the methods of rearing and slaughtering animals now being used. I might also want to know about the effect of a vegetarian diet on human health, and, considering the world food shortage, whether more or less food would be produced by giving up meat production. Once I have got evidence on these questions, I must assess it and bring it together with whatever moral views I hold. Depending on what method of moral reasoning I use, this may involve a calculation of which course of action produces greater happiness and less suffering; or it may mean an attempt to place myself in the positions of those affected by my decision; or it may lead me to attempt to "weigh up" conflicting duties and interests. Whatever method I employ, I must be aware of the possibility that my own desire to eat meat may lead to bias in my deliberations.

None of this procedure is easy—neither the gathering of information, nor the selection of what information is relevant, nor its combination with a basic moral position, nor the elimination of bias. Someone familiar with moral concepts and with moral arguments, who has ample time to gather information and think about it, may reasonably be expected to reach a soundly based conclusion more often than someone who is unfamiliar with moral concepts and moral arguments and has little time. So moral expertise would seem to be possible. The problem is not so much to know "the difference between right and wrong" as to decide what is right and what wrong.

If moral expertise is possible, have moral philosophers been right to disclaim it? Is the ordinary man just as likely to be expert in moral matters as the moral philosopher? On the basis of what has just been said, it would seem that the moral philosopher does have some important advantages over the ordinary man. First, his general training as a philosopher should make him more than ordinarily competent in argument and in the detection of invalid inferences. Next, his specific experience in moral philosophy gives him an understanding of moral concepts and of the logic of moral argument. The possibility of serious confusion arising if one engages in moral argument without a clear understanding of the concepts employed has been sufficiently emphasized in recent moral philosophy and does not need to be demonstrated here. Clarity is not an end in itself, but it is an aid to sound argument, and the need for clarity is something which moral philosophers have recognized. Finally, there is the simple fact that the moral philosopher can, if he wants, think full-time about moral issues, while most other people have some occupation to pursue which interferes with such reflection. It may sound silly to place much weight on this, but it is, I think very important. If we are to make moral judgments on some basis other than our unreflective intuitions, we need time, both for collecting facts and for thinking about them.

Moral philosophers have, then, certain advantages which could make them, relative to those who lack these advantages, experts in matters of morals. Of course, to be moral experts, it would be necessary for moral philosophers to do some fact-finding on whatever issue they were considering. Given a readiness to tackle normative issues, and to look at the relevant facts, it would be surprising if moral philosophers were not, in general, better suited to arrive at the right, or soundly based, moral conclusions than nonphilosophers. Indeed, if this were not the case, one might wonder whether moral philosophy was worthwhile.

About Ethics

FROM *Practical Ethics*

What Ethics Is Not

SOME PEOPLE THINK that morality is now out of date. They regard morality as a system of nasty puritanical prohibitions, mainly designed to stop people from having fun. Traditional moralists claim to be the defenders of morality in general, but they are really defending a particular moral code. They have been allowed to preempt the field to such an extent that when a newspaper headline reads BISHOP ATTACKS DECLINING MORAL STANDARDS, we expect to read yet again about promiscuity, homosexuality, pornography, and so on, and not about the puny amounts we give as overseas aid to poorer nations, or our reckless indifference to the natural environment of our planet.

So the first thing to say about ethics is that it is not a set of prohibitions particularly concerned with sex. Even in the era of AIDS, sex raises no unique moral issues at all. Decisions about sex may involve considerations of honesty, concern for others, prudence, and so on, but there is nothing special about sex in this respect, for the same could be said of decisions about driving a car. (In fact, the moral issues raised by driving a car, both from an environmental and from a safety point of view, are much more serious than those raised by sex.) Accordingly, this book contains no discussion of sexual morality. There are more important ethical issues to be considered.

Second, ethics is not an ideal system that is noble in theory but no good in practice. The reverse of this is closer to the truth: an ethical judgment that is no good in practice must suffer from a theoretical defect as well, for the whole point of ethical judgments is to guide practice.

Some people think that ethics is inapplicable to the real world because they regard it as a system of short and simple rules like "Do not lie," "Do not steal," and "Do not kill." It is not surprising that those who hold this view of ethics should also believe that ethics is not suited to life's complexities. In unusual situations, simple rules conflict; and even when they do not, following a rule can lead to disaster. It may normally be wrong to lie, but if you were living in Nazi Germany and the Gestapo came to your door looking for Jews, it would surely be right to deny the existence of the Jewish family hiding in your attic.

Like the failure of a restrictive sexual morality, the failure of an ethic of simple rules must not be taken as a failure of ethics as a whole. It is only a failure of one view of ethics, and not even an irremediable failure of that view. The deontologists—those who think that ethics is a system of rules—can rescue their position by finding more complicated and more specific rules that do not conflict with each other, or by ranking the rules in some hierarchical structure to resolve conflicts between them. Moreover, there is a long-standing approach to ethics that is quite untouched by the complexities that make simple rules difficult to apply. This is the consequentialist view. Consequentialists start not with moral rules but with goals. They assess actions by the extent to which they further these goals. The best-known, though not the only, consequentialist theory is utilitarianism. The classical utilitarian regards an action as right if it produces as much or more of an increase in the happiness of all affected by it than any alternative action, and wrong if it does not.

The consequences of an action vary according to the circumstances in which it is performed. Hence a utilitarian can never properly be accused of a lack of realism, or of a rigid adherence to ideals in defiance of practical experience. The utilitarian will judge lying bad in some circumstances and good in others, depending on its consequences.

Third, ethics is not something intelligible only in the context of religion. I shall treat ethics as entirely independent of religion.

Some theists say that ethics cannot do without religion because the very meaning of "good" is nothing other than "what God approves." Plato refuted a similar claim more than two thousand years ago by arguing that if the gods approve of some actions it must be because those actions are good, in which case it cannot be the gods' approval that makes them good. The alternative view makes divine approval entirely arbitrary: if the gods had happened to approve of torture and disapprove of helping our neighbors, torture would have been good and helping our neighbors bad. Some modern theists have attempted to extricate themselves from this type of dilemma by maintaining that God is good and so could not possibly approve of torture; but these theists are caught in a trap of their own making, for what can they possibly mean by the assertion that God is good? That God is approved of by God?

Traditionally, the more important link between religion and ethics was that religion was thought to provide a reason for doing what is right, the reason being that those who are virtuous will be rewarded by an eternity of bliss while the rest roast in hell. Not all religious thinkers have accepted this argument: Immanuel Kant, a most pious Christian, scorned anything that smacked of a self-interested motive for obeying the moral law. We must obey it, he said, for its own sake. Nor do we have to be Kantians to dispense with the motivation offered by traditional religion. There is a long line of thought that finds the source of ethics in the attitudes of benevolence and sympathy for others that most people have. This is, however, a complex topic, and since it is from [*Practical Ethics*] I shall not pursue it here. It is enough to say that our everyday observation of our fellow human beings clearly shows that ethical behavior does not require belief in heaven and hell.

The fourth, and last, claim about ethics that I shall deny in this chapter is that ethics is relative or subjective. At least, I shall deny these claims in some of the senses in which they are often made. This point requires a more extended discussion than the other three.

Let us take first the oft-asserted idea that ethics is relative to the society one happens to live in. This is true in one sense and false in another. It is true that, as we have already seen in discussing consequentialism, actions that are right in one situation because of their good consequences may be wrong in another situation because of their bad consequences. Thus casual sexual intercourse may be wrong when it leads to the existence of children who cannot be adequately cared for, and not wrong when, because of the existence of effective contraception, it does not lead to reproduction at all. But this is only a superficial form of relativism. While it suggests that the applicability of a specific principle like "Casual sex is wrong" may be relative to time and place, it says nothing against such a principle being objectively valid in specific circumstances, or against the universal applicability of a more general principle like "Do what increases happiness and reduces suffering."

The more fundamental form of relativism became popular in the nineteenth century when data on the moral beliefs and practices of far-flung societies began pouring in. To the strict reign of Victorian prudery the knowledge that there were places where sexual relations between unmarried people were regarded as perfectly wholesome brought the seeds of a revolution in sexual attitudes. It is not surprising that to some the new knowledge suggested, not merely that the moral code of nineteenth-century Europe was not objectively valid, but that no moral judgment can do more than reflect the customs of the society in which it is made.

Marxists adapted this form of relativism to their own theories. The ruling ideas of each period, they said, are the ideas of its ruling class, and so the morality of a society is relative to its dominant economic class, and thus indirectly relative to its economic basis. So they triumphantly refuted the claims of feudal and bourgeois morality to objective, universal validity. But this raises a problem: if all morality is relative, what is so special about communism? Why side with the proletariat rather than the bourgeoisie?

Engels dealt with this problem in the only way possible, by abandoning relativism in favor of the more limited claim that the morality of a society divided into classes will always be relative to the ruling class, although the morality of a society without class antagonisms could be a "really human" morality. This is no longer relativism at all. Nevertheless, Marxism, in a confused sort of way, still provides the impetus for a lot of woolly relativist ideas.

The problem that led Engels to abandon relativism defeats ordinary ethical relativism as well. Anyone who has thought through a difficult ethical decision knows that being told what our society thinks we ought to do does not settle the quandary. We have to reach our own decision. The beliefs and customs we were brought up with may exercise great influence on us, but once we start to reflect upon them we can decide whether to act in accordance with them, or to go against them.

The opposite view—that ethics is always relative to a particular society—has most implausible consequences. If our society disapproves of slavery, while another society approves of it, we have no basis to choose between these conflicting views. Indeed, on a relativist analysis there is really no conflict—when I say slavery is wrong I am really saying only that my society disapproves of slavery, and when the slave owners from the other society say that slavery is right, they are saying only that their society approves of it. Why argue? Obviously we could both be speaking the truth.

Worse still, the relativist cannot satisfactorily account for the nonconformist. If "slavery is wrong" means "my society disapproves of slavery," then someone who lives in a society that does not disapprove of slavery is, in claiming that slavery is wrong, making a simple factual error. An opinion poll could demonstrate the error of an ethical judgment. Would-be reformers are therefore in a parlous situation: when they set out to change the ethical views of their fellow-citizens, they are *necessarily* mistaken; it is only when they succeed in winning most of the society over to their own views that those views become right.

These difficulties are enough to sink ethical relativism; ethical subjectivism at least avoids making nonsense of the valiant efforts of would-be moral reformers, for it makes ethical judgments depend on the approval or disapproval of the person making the judgment, rather than that person's society. There are other difficulties, though, that at least some forms of ethical subjectivism cannot overcome.

If those who say that ethics is subjective mean by this that when I say that cruelty to animals is wrong I am really saying only that I disapprove of cruelty to animals, they are faced with an aggravated form of one of the difficulties of relativism: the inability to account for ethical disagreement. What for the relativist was true of disagreement between people from different societies is for the subjectivist true of disagreement between any two people. I say cruelty to animals is wrong: someone else says it is not wrong. If this means that I disapprove of cruelty to animals and someone else does not, both statements may be true and so there is nothing to argue about.

Other theories often described as "subjectivist" are not open to this objection. Suppose someone maintains that ethical judgments are neither true nor false because they do not describe anything—neither objective moral facts nor one's own subjective states of mind. This theory might hold that, as C. L. Stevenson suggested, ethical judgments express attitudes, rather than describe them, and we disagree about ethics because we try, by expressing our own attitude, to bring our listeners to a similar attitude. Or it might be, as R. M. Hare has urged, that ethical judgments are prescriptions and therefore more closely related to commands than to statements of fact. On this view we disagree because we care about what people do. Those features of ethical argument that imply the existence of objective moral standards can be explained away by maintaining that this is some kind of error—perhaps the legacy of the belief that ethics is a God-given system of law, or perhaps just another example of our tendency to objectify our personal wants and preferences. J. L. Mackie has defended this view.

Provided they are carefully distinguished from the crude form of subjectivism that sees ethical judgments as descriptions of the speaker's attitudes, these are plausible accounts of ethics. In their denial of a realm of ethical facts that is part of the real world, existing quite independently of us, they are no doubt correct; but does it follow from this that ethical judgments are immune from criticism, that there is no role for reason or argument in ethics, and that, from the standpoint of reason, any ethical judgment is as good as any other? I do not think it does, and none of the three philosophers referred to in the previous paragraph denies reason and argument a role in ethics, though they disagree as to the significance of this role.

This issue of the role that reason can play in ethics is the crucial point raised by the claim that ethics is subjective. The nonexistence of a mysterious realm of objective ethical facts does not imply the nonexistence of ethical reasoning. It may even help, since if we could arrive at ethical judgments only by intuiting these strange ethical facts, ethical argument would be more difficult still. So what has to be shown to put practical ethics on a sound basis is that ethical reasoning is possible. Here the temptation is to say simply that the proof of the pudding lies in the eating, and the proof that reasoning is possible in ethics is to be found in *Practical Ethics*, but this is not entirely satisfactory. From a theoretical point of view it is unsatisfactory because we might find ourselves reasoning about ethics without really understanding how this can happen; and from a practical point of view it is unsatisfactory because our reasoning is more likely to go astray if we lack a grasp of its foundations. I shall therefore attempt to say something about how we can reason in ethics.

What Ethics Is: One View

WHAT FOLLOWS IS a sketch of a view of ethics that allows reason an important role in ethical decisions. It is not the only possible view of ethics, but it is a plausible view. Once again, however, I shall have to pass over qualifications and objections worth a chapter to themselves. To those who think these undiscussed objections defeat the position I am advancing, I can only say, again, that this whole chapter may be treated as no more than a statement of the assumptions on which this book, *Practical Ethics*, is based. In that way it will at least assist in giving a clear view of what I take ethics to be.

What is it to make a moral judgment, or to argue about an ethical issue, or to live according to ethical standards? How do moral judgments differ from other practical judgments? Why do we regard a woman's decision to have an abortion as raising an ethical issue, but not her decision to change her job? What is the difference between a person who lives by ethical standards and one who doesn't?

All these questions are related, so we need to consider only one of them; but to do this we need to say something about the nature of ethics. Suppose that we have studied the lives of a number of different people, and we know a lot about what they do, what they believe, and so on. Can we then decide which of them are living by ethical standards and which are not?

We might think that the way to proceed here is to find out who believes it wrong to lie, cheat, steal, and so on and does not do any of these things, and who has no such beliefs and shows no such restraint in his actions. Then those in the first group would be living according to ethical standards and those in the second group would not be. But this procedure mistakenly assimilates two distinctions: the first is the distinction between living according to (what we judge to be) the right ethical standards and living according to (what we judge to be) mistaken ethical standards; the second is the distinction between living according to some ethical standards and living according to no ethical standards at all. Those who lie and cheat but do not believe what they are doing to be wrong may be living according to ethical standards. They may believe, for any of a number of possible reasons, that it is right to lie, cheat, steal, and so on. They are not living according to conventional ethical standards, but they may be living according to some other ethical standards.

This first attempt to distinguish the ethical from the nonethical was mistaken, but we can learn from our mistakes. We found that we must concede that those who hold unconventional ethical beliefs are still living according to ethical standards, *if they believe, for any reason, that it is right to do as they are doing.* The italicized condition gives us a clue to the answer we are seeking. The notion of living according to ethical standards is tied up with the notion of defending the way one is living, of giving a reason for it, of justifying it. Thus people may do all kinds of things we regard as wrong yet still be living according to ethical standards, if they are prepared to defend and justify what they do. We may find the justification inadequate, and may hold that the actions are wrong, but the attempt at justification, whether successful or not, is sufficient to bring the person's conduct within the domain of the ethical as opposed to the nonethical. When, on the other hand, people cannot put forward any justification for what they do, we may reject their claim to be living according to ethical standards, even if what they do is in accordance with conventional moral principles.

We can go further. If we are to accept that a person is living according to ethical standards, the justification must be of a certain kind. For instance, a justification in terms of self-interest alone will not do. When Macbeth, contemplating the murder of Duncan, admits that only "vaulting ambition" drives him to do it, he is admitting that the act cannot be justified ethically. "So that I can be king in his place" is not a weak attempt at an ethical justification for assassination; it is not the sort of reason that counts as an ethical justification at all. Self-interested acts must be shown to be compatible with more broadly based ethical principles if they are to be ethically defensible, for the notion of ethics carries with it the idea of something bigger than the individual. If I am to defend my conduct on ethical grounds, I cannot point only to the benefits it brings me. I must address myself to a larger audience.

From ancient times, philosophers and moralists have expressed the idea that ethical conduct is acceptable from a point of view that is somehow universal. The "Golden Rule" attributed to Moses, to be found in the book of Leviticus and subsequently repeated by Jesus, tells us to go beyond our own personal interests and "love thy neighbor as thyself"—in other words, give the same weight to the interests of others as one gives to one's own interests. The same idea of putting oneself in the position of another is involved in the other Christian formulation of the commandment, that we do to others as we would have them do to us. The Stoics held that ethics derives from a universal natural law. Kant developed this idea into his famous formula: "Act only on that maxim through which you can at the same time will that it should become a universal law." Kant's theory has itself been modified and developed by R. M. Hare, who sees universalizability as a logical feature of moral judgments. The eighteenth-century British philosophers Hutcheson, Hume, and Adam Smith appealed to an imaginary "impartial spectator" as the test of a moral judgment, and this theory has its modern version in the Ideal Observer theory. Utilitarians, from Jeremy Bentham to J.J.C. Smart, take it as axiomatic that in deciding moral issues "each counts for one and none for more than one"; while John Rawls, a leading contemporary critic of utilitarianism, incorporates essentially the same axiom into his own theory by deriving basic ethical principles from an imaginary choice in which those choosing do not know whether they will be the ones who gain or lose by the principles they select. Even continental European philosophers like the existentialist Jean-Paul Sartre and the critical theorist Jürgen Habermas, who differ in many ways from their English-speaking colleagues—and from each other—agree that ethics is in some sense universal.

One could argue endlessly about the merits of each of these characterizations of the ethical; but what they have in common is more important than their differences. They agree that an ethical principle cannot be justified in relation to any partial or sectional group. Ethics takes a universal point of view. This does not mean that a particular ethical judgment must be universally applicable. Circumstances alter cases, as we have seen. What it does mean is that in making ethical judgments we go beyond our own likes and dislikes. From an ethical point of view, the fact that it is I who benefit from, say, a more equal distribution of income, and you who lose by it, is irrelevant. Ethics requires us to go beyond "I" and "you" to the universal law, the universalizable judgment, the standpoint of the impartial spectator or ideal observer, or whatever we choose to call it.

Can we use this universal aspect of ethics to derive an ethical theory that will give us guidance about right and wrong? Philosophers from the Stoics to Hare and Rawls have attempted this. No attempt has met with general acceptance. The problem is that if we describe the universal aspect of ethics in bare, formal terms, a wide range of ethical theories, including quite irreconcilable ones, are compatible with this notion of universality; if, on the other hand, we build up our description of the universal aspect of ethics so that it leads us ineluctably to one particular ethical theory, we shall be accused of smuggling our own ethical beliefs into our definition of the ethical—and this definition was supposed to be broad enough, and neutral enough, to encompass all serious candidates for the status of "ethical theory." Since so many others have failed to overcome this obstacle to deducing an ethical theory from the universal aspect of ethics, it would be foolhardy to attempt to do so in a brief introduction to a work with a quite different aim. Nevertheless I shall propose something only a little less ambitious. The universal aspect of ethics, I suggest, does provide a persuasive, although not conclusive, reason for taking a broadly utilitarian position.

My reason for suggesting this is as follows. In accepting that ethical judgments must be made from a universal point of view, I am accepting that my own interests cannot, simply because they are my interests, count more than the interests of anyone else. Thus my very natural concern that my own interests be looked after must, when I think ethically, be extended to the interests of others. Now, imagine that I am trying to decide between two possible courses of action—perhaps whether to eat all the fruits I have collected myself, or to share them with others. Imagine, too, that I am deciding in a complete ethical vacuum, that I know nothing of any ethical considerations—I am, we might say, in a pre-ethical stage of thinking. How would I make up my mind? One thing that would be still relevant would be how the possible courses of action will affect my interests. Indeed, if we define "interests" broadly enough, so that we count anything people desire as in their interests (unless it is incompatible with another desire or desires), then it would seem that at this pre-ethical stage, *only* one's own interests can be relevant to the decision.

Suppose I then begin to think ethically, to the extent of recognizing that my own interests cannot count for more, simply because they are my own, than the interests of others. In place of my own interests, I now have to take into account the interests of all those affected by my decision. This requires me to weigh up all these interests and adopt the course of action most likely to maximize the interests of those affected. Thus at least at some level in my moral reasoning I must choose the course of action that has the best consequences, on balance, for all affected. (I say "at some level in my moral reasoning" because as we shall see later, there are utilitarian reasons for believing that we ought to try to calculate these consequences not for every ethical decision we make in our daily lives, but only in very unusual circumstances, or perhaps when we are reflecting on our choice of general principles to guide us in future. In other words, in the specific example given, at first glance one might think it obvious that sharing the fruits that I have gathered has better consequences for all affected than not sharing them. This may in the end also be the best general principle for us all to adopt, but before we can have grounds for believing this to be the case, we must also consider whether the effect of a general practice of sharing gathered fruits will benefit all those affected, by bringing about a more equal distribution, or whether it will reduce the amount of food gathered, because some will cease to gather anything if they know that they will get sufficient from their share of what others gather.)

The way of thinking I have outlined is a form of utilitarianism. It differs from classical utilitarianism in that "best consequences" is understood as meaning what, on balance, furthers the interests of those affected, rather than merely what increases pleasure and reduces pain. (It has, however, been suggested that classical utilitarians like Bentham and John Stuart Mill used "pleasure" and "pain" in a broad sense that allowed them to include achieving what one desired as a "pleasure" and the reverse as a "pain." If this interpretation is correct, the difference between classical utilitarianism and utilitarianism based on interests disappears.)

What does this show? It does not show that utilitarianism can be deduced from the universal aspect of ethics. There are other ethical ideals—like individual rights, the sanctity of life, justice, purity, and so on—that are universal in the required sense, and are, at least in some versions, incompatible with utilitarianism. It does show that we very swiftly arrive at an initially utilitarian position once we apply the universal aspect of ethics to simple, pre-ethical decision making. This, I believe, places the onus of proof on those who seek to go beyond utilitarianism. The utilitarian position is a minimal one, a first base that we reach by universalizing self-interested decision making. We cannot, if we are to think ethically, refuse to take this step. If we are to be persuaded that we should go beyond utilitarianism and accept nonutilitarian moral rules or ideals, we need to be provided with good reasons for taking this further step. Until such reasons are produced, we have some grounds for remaining utilitarians.

Across the Species Barrier

Preface to the 1975 Edition

FROM *Animal Liberation*

THIS BOOK, *Animal Liberation*, is about the tyranny of human over nonhuman animals. This tyranny has caused and today is still causing an amount of pain and suffering that can only be compared with that which resulted from the centuries of tyranny by white humans over black humans. The struggle against this tyranny is a struggle as important as any of the moral and social issues that have been fought over in recent years.

Most readers will take what they have just read to be a wild exaggeration. Five years ago I myself would have laughed at the statements I have now written in complete seriousness. Five years ago I did not know what I know today. If you read this book carefully, paying special attention to the second and third chapters, you will then know as much of what I know about the oppression of animals as it is possible to get into a book of reasonable length. Then you will be able to judge if my opening paragraph is a wild exaggeration or a sober estimate of a situation largely unknown to the general public. So I do not ask you to believe my opening paragraph now. All I ask is that you reserve your judgment until you have read the book.

SOON AFTER I began work on this book my wife and I were invited to tea—we were living in England at the time—by a lady who had heard that I was planning to write about animals. She herself was very interested in animals, she said, and she had a friend who had already written a book about animals and would be so keen to meet us.

When we arrived, our hostess's friend was already there, and she certainly was keen to talk about animals. "I do love animals," she began. "I have a dog and two cats, and do you know they get on together wonderfully well. Do you know Mrs. Scott? She runs a little hospital for sick pets . . ." and she was off. She paused while refreshments were served, took a ham sandwich, and then asked us what pets we had.

We told her we didn't own any pets. She looked a little surprised, and took a bite of her sandwich. Our hostess, who had now finished serving the sandwiches, joined us and took up the conversation: "But you *are* interested in animals, aren't you, Mr. Singer?"

We tried to explain that we were interested in the prevention of suffering and misery; that we were opposed to arbitrary discrimination; that we thought it wrong to inflict needless suffering on another being, even if that being was not a member of our own species; and that we believed animals were ruthlessly and cruelly exploited by humans, and we wanted this changed. Otherwise, we said, we were not especially "interested in" animals. Neither of us had ever been inordinately fond of dogs, cats, or horses in the way that many people are. We didn't "love" animals. We simply wanted them treated as the independent sentient beings that they are, and not as a means to human ends—as the pig whose flesh was now in our hostess's sandwiches had been treated.

This book is not about pets. It is not likely to be comfortable reading for those who think that love for animals involves no more than stroking a cat or feeding the birds in the garden. It is intended rather for people who are concerned about ending oppression and exploitation wherever they occur, and in seeing that the basic moral principle of equal consideration of interests is not arbitrarily restricted to members of our own species. The assumption that in order to be interested in such matters one must be an "animal-lover" is itself an indication of the absence of the slightest inkling that the moral standards that we apply among human beings might extend to other animals. No one, except a racist concerned to smear his opponents as "nigger-lovers," would suggest that in order to be concerned about equality for mistreated racial minorities you have to love those minorities, or regard them as cute and cuddly. So why make this assumption about people who work for improvements in the conditions of animals?

The portrayal of those who protest against cruelty to animals as sentimental, emotional "animal-lovers" has had the effect of excluding the entire issue of our treatment of nonhumans from serious political and moral discussion. It is easy to see why we do this. If we did give the issue serious consideration—if, for instance, we looked closely at the conditions in which animals live in the modern "factory farms" that produce our meat—we might be made uncomfortable about ham sandwiches, roast beef, fried chicken, and all those other items in our diet that we prefer not to think of as dead animals.

This book makes no sentimental appeals for sympathy toward "cute" animals. I am no more outraged by the slaughter of horses or dogs for meat than I am by the slaughter of pigs for this purpose. When the United States Defense Department finds that its use of beagles to test lethal gases has evoked a howl of protest and offers to use rats instead, I am not appeased.

This book is an attempt to think through, carefully and consistently, the question of how we ought to treat nonhuman animals. In the process it exposes the prejudices that lie behind our present attitudes and behavior. In the chapters that describe what these attitudes mean in practical terms—how animals suffer from the tyranny of human beings—there are passages that will arouse some emotions. These will, I hope, be emotions of anger and outrage, coupled with a determination to do something about the practices described. Nowhere in this book, however, do I appeal to the reader's emotions where they cannot be supported by reason. When there are unpleasant things to be described it would be dishonest to try to describe them in some neutral way that hid their real unpleasantness. You cannot write objectively about the experiments of the Nazi concentration camp "doctors" on those they considered "subhuman" without stirring emotions; and the same is true of a description of some of the experiments performed today on nonhumans in laboratories in America, Britain, and elsewhere. The ultimate justification for opposition to both these kinds of experiments, though, is not emotional. It is an appeal to basic moral principles which we all accept, and the application of these principles to the victims of both kinds of experiment is demanded by reason, not emotion.

The title of this book has a serious point behind it. A liberation movement is a demand for an end to prejudice and discrimination based on an arbitrary characteristic like race or sex. The classic instance is the black liberation movement. The immediate appeal of this movement, and its initial, if limited, success, made it a model for other oppressed groups. We soon became familiar with gay liberation and movements on behalf of American Indians and Spanish-speaking Americans. When a majority group—women—began their campaign, some thought we had come to the end of the road. Discrimination on the basis of sex, it was said, was the last form of discrimination to be universally accepted and practiced without secrecy or pretense, even in those liberal circles that have long prided themselves on their freedom from prejudice against racial minorities.

We should always be wary of talking of "the last remaining form of discrimination." If we have learned anything from the liberation movements, we should have learned how difficult it is to be aware of latent prejudices in our attitudes to particular groups until these prejudices are forcefully pointed out to us.

A liberation movement demands an expansion of our moral horizons. Practices that were previously regarded as natural and inevitable come to be seen as the result of an unjustifiable prejudice. Who can say with any confidence that none of his or her attitudes and practices can legitimately be questioned? If we wish to avoid being numbered among the oppressors, we must be prepared to rethink all our attitudes to other groups, including the most fundamental of them. We need to consider our attitudes from the point of view of those who suffer by them, and by the practices that follow from them. If we can make this unaccustomed mental switch, we may discover a pattern in our attitudes and practices that operates so as consistently to benefit the same group—usually the group to which we ourselves belong—at the expense of another group. So we come to see that there is a case for a new liberation movement.

The aim of this book is to lead you to make this mental switch in your attitudes and practices toward a very large group of beings: members of species other than our own. I believe that our present attitudes to these beings are based on a long history of prejudice and arbitrary discrimination. I argue that there can be no reason—except the selfish desire to preserve the privileges of the exploiting group—for refusing to extend the basic principle of equality of consideration to members of other species. I ask you to recognize that your attitudes to members of other species are a form of prejudice no less objectionable than prejudice about a person's race or sex.

In comparison with other liberation movements, animal liberation has a lot of handicaps. First and most obvious is the fact that members of the exploited group cannot themselves make an organized protest against the treatment they receive (though they can and do protest to the best of their abilities individually). We have to speak up on behalf of those who cannot speak for themselves. You can appreciate how serious this handicap is by asking yourself how long blacks would have had to wait for equal rights if they had not been able to stand up for themselves and demand those rights. The less able a group is to stand up and organize against oppression, the more easily it is oppressed.

More significant still for the prospects of the animal liberation movement is the fact that almost all of the oppressing group are directly involved in, and see themselves as benefiting from, the oppression. There are few humans indeed who can view the oppression of animals with the detachment possessed, say, by northern whites debating the institution of slavery in the southern states of the Union. People who eat pieces of slaughtered nonhumans every day find it hard to believe that they are doing wrong; and they also find it hard to imagine what else they could eat. On this issue, anyone who eats meat is an interested party. Meat eaters benefit—or at least they think they benefit—from the present disregard of the interests of nonhuman animals. This makes persuasion more difficult. How many southern slaveholders were persuaded by the arguments used by the northern abolitionists and accepted by nearly all of us today? Some, but not many. I can and do ask you to put aside your interest in eating meat when considering the arguments of this book; but I know from my own experience that with the best will in the world this is not an easy thing to do. For behind the mere momentary desire to eat meat on a particular occasion lie many years of habitual meat-eating which have conditioned our attitudes to animals.

Habit. That is the final barrier that the animal liberation movement faces. Habits not only of diet but also of thought and language must be challenged and altered. Habits of thought lead us to brush aside descriptions of cruelty to animals as emotional, for "animal-lovers only"; or if not that, then anyway the problem is so trivial in comparison with the problems of human beings that no sensible person could give it time and attention. This too is a prejudice—for how can one know that a problem is trivial until one has taken the time to examine its extent? Although in order to allow a more thorough treatment this book deals with only two of the many areas in which humans cause other animals to suffer, I do not think anyone who reads it to the end will ever again think that the only problems that merit time and energy are problems concerning humans.

The habits of thought that lead us to disregard the interests of animals can be challenged, as they are challenged in the following pages. This challenge has to be expressed in a language, which in this case happens to be English. The English language, like other languages, reflects the prejudices of its users. So authors who wish to challenge these prejudices are in a well-known type of bind: either they use language that reinforces the very prejudices they wish to challenge, or else they fail to communicate with their audience. This book has already been forced along the former of these paths. We commonly use the word "animal" to mean "animals other than human beings." This usage sets humans apart from other animals, implying that we are not ourselves animals—an implication that everyone who has had elementary lessons in biology knows to be false.

In the popular mind the term "animal" lumps together beings as different as oysters and chimpanzees, while placing a gulf between chimpanzees and humans, although our relationship to those apes is much closer than the oyster's. Since there exists no other short term for the nonhuman animals, I have, in the title of this book and elsewhere in these pages, had to use "animal" as if it did not include the human animal. This is a regrettable lapse from the standards of revolutionary purity, but it seems necessary for effective communication. Occasionally, however, to remind you that this is a matter of convenience only, I shall use longer, more accurate modes of referring to what was once called "the brute creation." In other cases, too, I have tried to avoid language which tends to degrade animals or disguise the nature of the food we eat.

The basic principles of animal liberation are very simple. I have tried to write a book that is clear and easy to understand, requiring no expertise of any kind. It is necessary, however, to begin with a discussion of the principles that underlie what I have to say. While there should be nothing here that is difficult, readers unused to this kind of discussion might find the first chapter rather abstract. Don't be put off. In the next chapters we get down to the little-known details of how our species oppresses others under our control. There is nothing abstract about this oppression, or about the chapters that describe it.

If the recommendations made in the following chapters are accepted, millions of animals will be spared considerable pain. Moreover, millions of humans will benefit too. As I write, people are starving to death in many parts of the world, and many more are in imminent danger of starvation. The United States government has said that because of poor harvests and diminished stocks of grain it can provide only limited—and inadequate—assistance; but the heavy emphasis in affluent nations on rearing animals for food wastes several times as much food as it produces. By ceasing to rear and kill animals for food, we can make so much extra food available for humans that, properly distributed, it would eliminate starvation and malnutrition from this planet. Animal liberation is human liberation, too.

All Animals Are Equal . . .

FROM *Animal Liberation*

or why the ethical principle on which human equality rests requires us to extend equal consideration to animals too

ANIMAL LIBERATION may sound more like a parody of other liberation movements than a serious objective. The idea of "The Rights of Animals" actually was once used to parody the case for women's rights. When Mary Wollstonecraft, a forerunner of today's feminists, published her *Vindication of the Rights of Woman* in 1792, her views were widely regarded as absurd, and before long an anonymous publication appeared entitled *A Vindication of the Rights of Brutes*. The author of this satirical work (now known to have been Thomas Taylor, a distinguished Cambridge philosopher) tried to refute Mary Wollstonecraft's arguments by showing that they could be carried one stage further. If the argument for equality was sound when applied to women, why should it not be applied to dogs, cats, and horses? The reasoning seemed to hold for these "brutes" too; yet to hold that brutes had rights was manifestly absurd. Therefore the reasoning by which this conclusion had been reached must be unsound, and if unsound when applied to brutes, it must also be unsound when applied to women, since the very same arguments had been used in each case.

In order to explain the basis of the case for the equality of animals, it will be helpful to start with an examination of the case for the equality of women. Let us assume that we wish to defend the case for women's rights against the attack by Thomas Taylor. How should we reply?

One way in which we might reply is by saying that the case for equality between men and women cannot validly be extended to nonhuman animals. Women have a right to vote, for instance, because they are just as capable of making rational decisions about the future as men are; dogs, on the other hand, are incapable of understanding the significance of voting, so they cannot have the right to vote. There are many other obvious ways in which men and women resemble each other closely, while humans and animals differ greatly. So, it might be said, men and women are similar beings and should have similar rights, while humans and nonhumans are different and should not have equal rights.

The reasoning behind this reply to Taylor's analogy is correct up to a point, but it does not go far enough. There are obviously important differences between humans and other animals, and these differences must give rise to some differences in the rights that each have. Recognizing this evident fact, however, is no barrier to the case for extending the basic principle of equality to nonhuman animals. The differences that exist between men and women are equally undeniable, and the supporters of women's liberation are aware that these differences may give rise to different rights. Many feminists hold that women have the right to an abortion on request. It does not follow that since these same feminists are campaigning for equality between men and women, they must support the right of men to have abortions too. Since a man cannot have an abortion, it is meaningless to talk of his right to have one. Since dogs can't vote, it is meaningless to talk of their right to vote. There is no reason why either women's liberation or animal liberation should get involved in such nonsense. The extension of the basic principle of equality from one group to another does not imply that we must treat both groups in exactly the same way or grant exactly the same rights to both groups. Whether we should do so will depend on the nature of the members of the two groups. The basic principle of equality does not require equal or identical *treatment*; it requires equal consideration. Equal consideration for different beings may lead to different treatment and different rights.

So there is a different way of replying to Taylor's attempt to parody the case for women's rights, a way that does not deny the obvious differences between human beings and nonhumans but goes more deeply into the question of equality and concludes by finding nothing absurd in the idea that the basic principle of equality applies to so-called brutes. At this point such a conclusion may appear odd; but if we examine more deeply the basis on which our opposition to discrimination on grounds of race or sex ultimately rests, we will see that we would be on shaky ground if we were to demand equality for blacks, women, and other groups of oppressed humans while denying equal consideration to nonhumans. To make this clear we need to see, first, exactly why racism and sexism are wrong. When we say that all human beings, whatever their race, creed, or sex, are equal, what is it that we are asserting? Those who wish to defend hierarchical, inegalitarian societies have often pointed out that by whatever test we choose it simply is not true that all humans are equal. Like it or not we must face the fact that humans come in different shapes and sizes; they come with different moral capacities, different intellectual abilities, different amounts of benevolent feeling and sensitivity to the needs of others, different abilities to communicate effectively, and different capacities to experience pleasure and pain. In short, if the demand for equality were based on the actual equality of all human beings, we would have to stop demanding equality.

Still, one might cling to the view that the demand for equality among human beings is based on the actual equality of the different races and sexes. Although, it may be said, humans differ as individuals, there are no differences between the races and sexes as such. From the mere fact that a person is black or a woman we cannot infer anything about that person's intellectual or moral capacities. This, it may be said, is why racism and sexism are wrong. The white racist claims that whites are superior to blacks, but this is false; although there are differences among individuals, some blacks are superior to some whites in all of the capacities and abilities that could conceivably be relevant. The opponent of sexism would say the same: a person's sex is no guide to his or her abilities, and this is why it is unjustifiable to discriminate on the basis of sex.

The existence of individual variations that cut across the lines of race or sex, however, provides us with no defense at all against a more sophisticated opponent of equality, one who proposes that, say, the interests of all those with IQ scores below 100 be given less consideration than the interests of those with ratings over 100. Perhaps those scoring below the mark would, in this society, be made the slaves of those scoring higher. Would a hierarchical society of this sort really be so much better than one based on race or sex? I think not. But if we tie the moral principle of equality to the factual equality of the different races or sexes, taken as a whole, our opposition to racism and sexism does not provide us with any basis for objecting to this kind of inegalitarianism.

There is a second important reason why we ought not to base our opposition to racism and sexism on any kind of factual equality, even the limited kind that asserts that variations in capacities and abilities are spread evenly among the different races and between the sexes: we can have no absolute guarantee that these capacities and abilities really are distributed evenly, without regard to race or sex, among human beings. So far as actual abilities are concerned there do seem to be certain measurable differences both among races and between sexes. These differences do not, of course, appear in every case; they appear only when averages are taken. More important still, we do not yet know how many of these differences are really due to the different genetic endowments of the different races and sexes, and how many are due to poor schools, poor housing, and other factors that are the result of past and continuing discrimination. Perhaps all of the important differences will eventually prove to be environmental rather than genetic. Anyone opposed to racism and sexism will certainly hope that this will be so, for it will make the task of ending discrimination a lot easier; nevertheless, it would be dangerous to rest the case against racism and sexism on the belief that all significant differences are environmental in origin. The opponent of, say, racism who takes this line will be unable to avoid conceding that if differences in ability did after all prove to have some genetic connection with race, racism would in some way be defensible.

Fortunately there is no need to pin the case for equality on one particular outcome of a scientific investigation. The appropriate response to those who claim to have found evidence of genetically based differences in ability among the races or between the sexes is not to stick to the belief that the genetic explanation must be wrong, whatever evidence to the contrary may turn up; instead we should make it quite clear that the claim to equality does not depend on intelligence, moral capacity, physical strength, or similar matters of fact. Equality is a moral idea, not an assertion of fact. There is no logically compelling reason for assuming that a factual difference in ability between two people justifies any difference in the amount of consideration we give to their needs and interests. *The principle of the equality of human beings is not a description of an alleged actual equality among humans: it is a prescription of how we should treat human beings.*

Jeremy Bentham, the founder of the reforming utilitarian school of moral philosophy, incorporated the essential basis of moral equality into his system of ethics by means of the formula: "Each to count for one and none for more than one." In other words, the interests of every being affected by an action are to be taken into account and given the same weight as the like interests of any other being. A later utilitarian, Henry Sidgwick, put the point in this way: "The good of any one individual is of no more importance, from the point of view (if I may say so) of the Universe, than the good of any other." More recently the leading figures in contemporary moral philosophy have shown a great deal of agreement in specifying as a fundamental presupposition of their moral theories some similar requirement that works to give everyone's interests equal consideration—although these writers generally cannot agree on how this requirement is best formulated.[1]

It is an implication of this principle of equality that our concern for others and our readiness to consider their interests ought not to depend on what they are like or on what abilities they may possess. Precisely what our concern or consideration requires us to do may vary according to the characteristics of those affected by what we do: concern for the well-being of children growing up in America would require that we teach them to read; concern for the well-being of pigs may require no more than that we leave them with other pigs in a place where there is adequate food and room to run freely. But the basic element—the taking into account of the interests of the being, whatever those interests may be—must, according to the principle of equality, be extended to all beings, black or white, masculine or feminine, human or nonhuman.

Thomas Jefferson, who was responsible for writing the principle of the equality of men into the American Declaration of Independence, saw this point. It led him to oppose slavery even though he was unable to free himself fully from his slaveholding background. He wrote in a letter to the author of a book that emphasized the notable intellectual achievements of Negroes in order to refute the then common view that they had limited intellectual capacities:

> Be assured that no person living wishes more sincerely than I do, to see a complete refutation of the doubts I myself have entertained and expressed on the grade of understanding allotted to them by nature, and to find that they are on a par with ourselves . . . but whatever be their degree of talent it is no measure of their rights. Because Sir Isaac Newton was superior to others in understanding, he was not therefore lord of the property or persons of others.[2]

Similarly, when in the 1850s the call for women's rights was raised in the United States, a remarkable black feminist named Sojourner Truth made the same point in more robust terms at a feminist convention:

> They talk about this thing in the head; what do they call it? ["Intellect," whispered someone nearby.] That's it. What's that got to do with women's rights or Negroes' rights? If my cup won't hold but a pint and yours holds a quart, wouldn't you be mean not to let me have my little half-measure full?[3]

It is on this basis that the case against racism and the case against sexism must both ultimately rest; and it is in accordance with this principle that the attitude that we may call "speciesism," by analogy with racism, must also be condemned. Speciesism—the word is not an attractive one, but I can think of no better term—is a prejudice or attitude of bias in favor of the interests of members of one's own species and against those of members of other species. It should be obvious that the fundamental objections to racism and sexism made by Thomas Jefferson and Sojourner Truth apply equally to speciesism. If possessing a higher degree of intelligence does not entitle one human to use another for his or her own ends, how can it entitle humans to exploit nonhumans for the same purpose?[4]

Many philosophers and other writers have proposed the principle of equal consideration of interests, in some form or other, as a basic moral principle; but not many of them have recognized that this principle applies to members of other species as well as to our own. Jeremy Bentham was one of the few who did realize this. In a forward-looking passage written at a time when black slaves had been freed by the French but in the British dominions were still being treated in the way we now treat animals, Bentham wrote:

> The day *may* come when the rest of the animal creation may acquire those rights which never could have been withholden from them but by the hand of tyranny. The French have already discovered that the blackness of the skin is no reason why a human being should be abandoned without redress to the caprice of a tormentor. It may one day come to be recognized that the number of the legs, the villosity of the skin, or the termination of the *os sacrum* are reasons equally insufficient for abandoning a sensitive being to the same fate. What else is it that should trace the insuperable line? Is it the faculty of reason, or perhaps the faculty of discourse? But a full-grown horse or dog is beyond comparison a more rational, as well as a more conversable animal, than an infant of a day or a week or even a month, old. But suppose they were otherwise, what would it avail? The question is not, Can they *reason?* nor Can they *talk*? but, Can they *suffer*?[5]

In this passage Bentham points to the capacity for suffering as the vital characteristic that gives a being the right to equal consideration. The capacity for suffering—or more strictly, for suffering and/or enjoyment or happiness—is not just another characteristic like the capacity for language or higher mathematics. Bentham is not saying that those who try to mark "the insuperable line" that determines whether the interests of a being should be considered happen to have chosen the wrong characteristic. By saying that we must consider the interests of all beings with the capacity for suffering or enjoyment Bentham does not arbitrarily exclude from consideration any interests at all—as those who draw the line with reference to the possession of reason or language do. The capacity for suffering and enjoyment is *a prerequisite for having interests at all*, a condition that must be satisfied before we can speak of interests in a meaningful way. It would be nonsense to say that it was not in the interests of a stone to be kicked along the road by a schoolboy. A stone does not have interests because it cannot suffer. Nothing that we can do to it could possibly make any difference to its welfare. The capacity for suffering and enjoyment is, however, not only necessary, but also sufficient for us to say that a being has interests—at an absolute minimum, an interest in not suffering. A mouse, for example, does have an interest in not being kicked along the road, because it will suffer if it is.

Although Bentham speaks of "rights" in the passage I have quoted, the argument is really about equality rather than about rights. Indeed, in a different passage, Bentham famously described "natural rights" as "nonsense" and "natural and imprescriptable rights" as "nonsense upon stilts." He talked of moral rights as a shorthand way of referring to protections that people and animals morally ought to have; but the real weight of the moral argument does not rest on the assertion of the existence of the right, for this in turn has to be justified on the basis of the possibilities for suffering and happiness. In this way we can argue for equality for animals without getting embroiled in philosophical controversies about the ultimate nature of rights.

In misguided attempts to refute the arguments of this book, some philosophers have gone to much trouble developing arguments to show that animals do not have rights.[6] They have claimed that to have rights a being must be autonomous, or must be a member of a community, or must have the ability to respect the rights of others, or must possess a sense of justice. These claims are irrelevant to the case for animal liberation. The language of rights is a convenient political shorthand. It is even more valuable in the era of thirty-second TV news clips than it was in Bentham's day; but in the argument for a radical change in our attitude to animals, it is in no way necessary.

If a being suffers there can be no moral justification for refusing to take that suffering into consideration. No matter what the nature of the being, the principle of equality requires that its suffering be counted equally with the like suffering—insofar as rough comparisons can be made—of any other being. If a being is not capable of suffering, or of experiencing enjoyment or happiness, there is nothing to be taken into account. So the limit of sentience (using the term as a convenient if not strictly accurate shorthand for the capacity to suffer and/or experience enjoyment) is the only defensible boundary of concern for the interests of others. To mark this boundary by some other characteristic like intelligence or rationality would be to mark it in an arbitrary manner. Why not choose some other characteristic, like skin color?

Racists violate the principle of equality by giving greater weight to the interests of members of their own race when there is a clash between their interests and the interests of those of another race. Sexists violate the principle of equality by favoring the interests of their own sex. Similarly, speciesists allow the interests of their own species to override the greater interests of members of other species. The pattern is identical in each case.

MOST HUMAN BEINGS are speciesists. The following chapters show that ordinary human beings—not a few exceptionally cruel or heartless humans, but the overwhelming majority of humans—take an active part in, acquiesce in, and allow their taxes to pay for practices that require the sacrifice of the most important interests of members of other species in order to promote the most trivial interests of our own species.

There is, however, one general defense of the practices to be described in the next two chapters that needs to be disposed of before we discuss the practices themselves. It is a defense which, if true, would allow us to do anything at all to nonhumans for the slightest reason, or for no reason at all, without incurring any justifiable reproach. This defense claims that we are never guilty of neglecting the interests of other animals for one breathtakingly simple reason: they have no interests. Nonhuman animals have no interests, according to this view, because they are not capable of suffering. By this is not meant merely that they are not capable of suffering in all the ways that human beings are—for instance, that a calf is not capable of suffering from the knowledge that it will be killed in six months' time. That modest claim is, no doubt, true; but it does not clear humans of the charge of speciesism, since it allows that animals may suffer in other ways—for instance, by being given electric shocks, or being kept in small, cramped cages. The defense I am about to discuss is the much more sweeping, although correspondingly less plausible, claim that animals are incapable of suffering in any way at all; that they are, in fact, unconscious automata, possessing neither thoughts nor feelings nor a mental life of any kind.

Although, as we shall see in a later chapter, the view that animals are automata was proposed by the seventeenth-century French philosopher René Descartes, to most people, then and now, it is obvious that if, for example, we stick a sharp knife into the stomach of an unanesthetized dog, the dog will feel pain. That this is so is assumed by the laws in most civilized countries that prohibit wanton cruelty to animals. Readers whose common sense tells them that animals do suffer may prefer to skip the remainder of this section, moving straight on to page 41, since the pages in between do nothing but refute a position that they do not hold. Implausible as it is, though, for the sake of completeness this skeptical position must be discussed.

Do animals other than humans feel pain? How do we know? Well, how do we know if anyone, human or nonhuman, feels pain? We know that we ourselves can feel pain. We know this from the direct experience of pain that we have when, for instance, somebody presses a lighted cigarette against the back of our hand. But how do we know that anyone else feels pain? We cannot directly experience anyone else's pain, whether that "anyone" is our best friend or a stray dog. Pain is a state of consciousness, a "mental event," and as such it can never be observed. Behavior like writhing, screaming, or drawing one's hand away from the lighted cigarette is not pain itself; nor are the recordings a neurologist might make of activity within the brain observations of pain itself. Pain is something that we feel, and we can only infer that others are feeling it from various external indications.

In theory, we *could* always be mistaken when we assume that other human beings feel pain. It is conceivable that one of our close friends is really a cleverly constructed robot, controlled by a brilliant scientist so as to give all the signs of feeling pain, but really no more sensitive than any other machine. We can never know, with absolute certainty, that this is not the case. But while this might present a puzzle for philosophers, none of us has the slightest real doubt that our close friends feel pain just as we do. This is an inference, but a perfectly reasonable one, based on observations of their behavior in situations in which we would feel pain, and on the fact that we have every reason to assume that our friends are beings like us, with nervous systems like ours that can be assumed to function as ours do and to produce similar feelings in similar circumstances.

If it is justifiable to assume that other human beings feel pain as we do, is there any reason why a similar inference should be unjustifiable in the case of other animals?

Nearly all the external signs that lead us to infer pain in other humans can be seen in other species, especially the species most closely related to us—the species of mammals and birds. The behavioral signs include writhing, facial contortions, moaning, yelping or other forms of calling, attempts to avoid the source of pain, appearance of fear at the prospect of its repetition, and so on. In addition, we know that these animals have nervous systems very like ours, which respond physiologically as ours do when the animal is in circumstances in which we would feel pain: an initial rise of blood pressure, dilated pupils, perspiration, an increased pulse rate, and, if the stimulus continues, a fall in blood pressure. Although human beings have a more developed cerebral cortex than other animals, this part of the brain is concerned with thinking functions rather than with basic impulses, emotions, and feelings. These impulses, emotions, and feelings are located in the diencephalon, which is well developed in many other species of animals, especially mammals and birds.[7]

We also know that the nervous systems of other animals were not artificially constructed—as a robot might be artificially constructed—to mimic the pain behavior of humans. The nervous systems of animals evolved as our own did, and in fact the evolutionary history of human beings and other animals, especially mammals, did not diverge until the central features of our nervous systems were already in existence. A capacity to feel pain obviously enhances a species' prospects of survival, since it causes members of the species to avoid sources of injury. It is surely unreasonable to suppose that nervous systems that are virtually identical physiologically, have a common origin and a common evolutionary function, and result in similar forms of behavior in similar circumstances should actually operate in an entirely different manner on the level of subjective feelings.

It has long been accepted as sound policy in science to search for the simplest possible explanation of whatever it is we are trying to explain.

Occasionally it has been claimed that it is for this reason "unscientific" to explain the behavior of animals by theories that refer to the animal's conscious feelings, desires, and so on—the idea being that if the behavior in question can be explained without invoking consciousness or feelings, that will be the simpler theory. Yet we can now see that such explanations, when assessed with respect to the actual behavior of both human and nonhuman animals, are actually far more complex than rival explanations. For we know from our own experience that explanations of our own behavior that did not refer to consciousness and the feeling of pain would be incomplete; and it is simpler to assume that the similar behavior of animals with similar nervous systems is to be explained in the same way than to try to invent some other explanation for the behavior of nonhuman animals as well as an explanation for the divergence between humans and nonhumans in this respect.

The overwhelming majority of scientists who have addressed themselves to this question agree. Lord Brain, one of the most eminent neurologists of our time, has said:

> I personally can see no reason for conceding mind to my fellow men and denying it to animals. . . . I at least cannot doubt that the interests and activities of animals are correlated with awareness and feeling in the same way as my own, and which may be, for aught I know, just as vivid.[8]

The author of a book on pain writes:

> Every particle of factual evidence supports the contention that the higher mammalian vertebrates experience pain sensations at least as acute as our own. To say that they feel less because they are lower animals is an absurdity; it can easily be shown that many of their senses are far more acute than ours—visual acuity in certain birds, hearing in most wild animals, and touch in others; these animals depend more than we do today on the sharpest possible awareness of a hostile environment. Apart from the complexity of the cerebral cortex (which does not directly perceive pain) their nervous systems are almost identical to ours and their reactions to pain remarkably similar, though lacking (so far as we know) the philosophical and moral overtones. The emotional element is all too evident, mainly in the form of fear and anger.[9]

In Britain, three separate expert government committees on matters relating to animals have accepted the conclusion that animals feel pain.

After noting the obvious behavioral evidence for this view, the members of the Committee on Cruelty to Wild Animals, set up in 1951, said:

> . . . we believe that the physiological, and more particularly the anatomical, evidence fully justifies and reinforces the commonsense belief that animals feel pain.

And after discussing the evolutionary value of pain the committee's report concluded that pain is "of clear-cut biological usefulness" and this is "a third type of evidence that animals feel pain." The committee members then went on to consider forms of suffering other than mere physical pain and added that they were "satisfied that animals do suffer from acute fear and terror." Subsequent reports by British government committees on experiments on animals and on the welfare of animals under intensive farming methods agreed with this view, concluding that animals are capable of suffering both from straightforward physical injuries and from fear, anxiety, stress, and so on.[10] Finally, within the last decade, the publication of scientific studies with titles such as *Animal Thought*, *Animal Thinking*, and *Animal Suffering: The Science of Animal Welfare* have made it plain that conscious awareness in nonhuman animals is now generally accepted as a serious subject for investigation.[11]

That might well be thought enough to settle the matter, but one more objection needs to be considered. Human beings in pain, after all, have one behavioral sign that nonhuman animals do not have: a developed language. Other animals may communicate with each other, but not, it seems, in the complicated way we do. Some philosophers, including Descartes, have thought it important that while humans can tell each other about their experience of pain in great detail, other animals cannot. (Interestingly, this once neat dividing line between humans and other species has now been threatened by the discovery that chimpanzees can be taught a language.[12]) But as Bentham pointed out long ago, the ability to use language is not relevant to the question of how a being ought to be treated—unless that ability can be linked to the capacity to suffer, so that the absence of a language casts doubt on the existence of this capacity.

This link may be attempted in two ways. First, there is a hazy line of philosophical thought, deriving perhaps from some doctrines associated with the influential philosopher Ludwig Wittgenstein, which maintains that we cannot meaningfully attribute states of consciousness to beings without language. This position seems to me very implausible. Language may be necessary for abstract thought, at some level anyway; but states like pain are more primitive and have nothing to do with language.

The second and more easily understood way of linking language and the existence of pain is to say that the best evidence we can have that other creatures are in pain is that they tell us that they are. This is a distinct line of argument, for it is denying not that non-language-users conceivably *could* suffer, but only that we could ever have sufficient reason to *believe* that they are suffering. Still, this line of argument fails too. As Jane Goodall has pointed out in her study of chimpanzees, *In the Shadow of Man*, when it comes to the expression of feelings and emotions language is less important than nonlinguistic modes of communication such as a cheering pat on the back, an exuberant embrace, a clasp of the hands, and so on. The basic signals we use to convey pain, fear, anger, love, joy, surprise, sexual arousal, and many other emotional states are not specific to our own species.[13] The statement "I am in pain" may be one piece of evidence for the conclusion that the speaker is in pain, but it is not the only possible evidence, and since people sometimes tell lies, not even the best possible evidence.

Even if there were stronger grounds for refusing to attribute pain to those who do not have a language, the consequences of this refusal might lead us to reject the conclusion. Human infants and young children are unable to use language. Are we to deny that a year-old child can suffer? If not, language cannot be crucial. Of course, most parents understand the responses of their children better than they understand the responses of other animals; but this is just a fact about the relatively greater knowledge that we have of our own species and the greater contact we have with infants as compared with animals. Those who have studied the behavior of other animals and those who have animals as companions soon learn to understand their responses as well as we understand those of an infant, and sometimes better.

So to conclude: there are no good reasons, scientific or philosophical, for denying that animals feel pain. If we do not doubt that other humans feel pain, we should not doubt that other animals do so too.

Animals can feel pain. As we saw earlier, there can be no moral justification for regarding the pain (or pleasure) that animals feel as less important than the same amount of pain (or pleasure) felt by humans. But what practical consequences follow from this conclusion? To prevent misunderstanding I shall spell out what I mean a little more fully.

If I give a horse a hard slap across its rump with my open hand, the horse may start, but it presumably feels little pain. Its skin is thick enough to protect it against a mere slap. If I slap a baby in the same way, however, the baby will cry and presumably feel pain, for its skin is more sensitive. So it is worse to slap a baby than a horse, if both slaps are administered with equal force. But there must be some kind of blow—I don't know exactly what it would be, but perhaps a blow with a heavy stick—that would cause the horse as much pain as we cause a baby by slapping it with our hand. That is what I mean by "the same amount of pain," and if we consider it wrong to inflict that much pain on a baby for no good reason, then we must, unless we are speciesists, consider it equally wrong to inflict the same amount of pain on a horse for no good reason.

Other differences between humans and animals cause other complications. Normal adult human beings have mental capacities that will, in certain circumstances, lead them to suffer more than animals would in the same circumstances. If, for instance, we decided to perform extremely painful or lethal scientific experiments on normal adult humans, kidnapped at random from public parks for this purpose, adults who enjoy strolling in parks would become fearful that they would be kidnapped. The resultant terror would be a form of suffering additional to the pain of the experiment. The same experiments performed on nonhuman animals would cause less suffering, since the animals would not have the anticipatory dread of being kidnapped and experimented upon. This does not mean, of course, that it would be *right* to perform the experiment on animals, but only that there is a reason, which is *not* speciesist, for preferring to use animals rather than normal adult human beings, if the experiment is to be done at all. It should be noted, however, that this same argument gives us a reason for preferring to use human infants—orphans perhaps—or severely retarded human beings for experiments, rather than adults, since infants and retarded humans would also have no idea of what was going to happen to them. So far as this argument is concerned nonhuman animals and infants and retarded humans are in the same category; and if we use this argument to justify experiments on nonhuman animals, we have to ask ourselves whether we are also prepared to allow experiments on human infants and retarded adults; and if we make a distinction between animals and these humans, on what basis can we do it, other than a bare-faced—and morally indefensible—preference for members of our own species?

There are many matters in which the superior mental powers of normal adult humans make a difference: anticipation, more detailed memory, greater knowledge of what is happening, and so on. Yet these differences do not all point to greater suffering on the part of the normal human being. Sometimes animals may suffer more because of their more limited understanding. If, for instance, we are taking prisoners in wartime, we can explain to them that although they must submit to capture, search, and confinement, they will not otherwise be harmed and will be set free at the conclusion of hostilities. If we capture wild animals, however, we cannot explain that we are not threatening their lives. A wild animal cannot distinguish an attempt to overpower and confine from an attempt to kill; the one causes as much terror as the other.

It may be objected that comparisons of the sufferings of different species are impossible to make and that for this reason when the interests of animals and humans clash the principle of equality gives no guidance. It is probably true that comparisons of suffering between members of different species cannot be made precisely, but precision is not essential. Even if we were to prevent the infliction of suffering on animals only when it is quite certain that the interests of humans will not be affected to anything like the extent that animals are affected, we would be forced to make radical changes in our treatment of animals that would involve our diet; the farming methods we use; experimental procedures in many fields of science; our approach to wildlife and to hunting, trapping, and the wearing of furs; and areas of entertainment like circuses, rodeos, and zoos. As a result, a vast amount of suffering would be avoided.

So far I have said a lot about inflicting suffering on animals, but nothing about killing them. This omission has been deliberate. The application of the principle of equality to the infliction of suffering is, in theory at least, fairly straightforward. Pain and suffering are in themselves bad and should be prevented or minimized, irrespective of the race, sex, or species of the being that suffers. How bad a pain is depends on how intense it is and how long it lasts, but pains of the same intensity and duration are equally bad, whether felt by humans or animals.

The wrongness of killing a being is more complicated. I have kept, and shall continue to keep, the question of killing in the background because in the present state of human tyranny over other species the more simple, straightforward principle of equal consideration of pain or pleasure is a sufficient basis for identifying and protesting against all the major abuses of animals that human beings practice. Nevertheless, it is necessary to say something about killing.

Just as most human beings are speciesists in their readiness to cause pain to animals when they would not cause a similar pain to humans for the same reason, so most human beings are speciesists in their readiness to kill other animals when they would not kill human beings. We need to proceed more cautiously here, however, because people hold widely differing views about when it is legitimate to kill humans, as the continuing debates over abortion and euthanasia attest. Nor have moral philosophers been able to agree on exactly what it is that makes it wrong to kill human beings, and under what circumstances killing a human being may be justifiable.

Let us consider first the view that it is always wrong to take an innocent human life. We may call this the "sanctity of life" view. People who take this view oppose abortion and euthanasia. They do not usually, however, oppose the killing of nonhuman animals—so perhaps it would be more accurate to describe this view as the "sanctity of *human* life" view. The belief that human life, and only human life, is sacrosanct is a form of speciesism. To see this, consider the following example.

Assume that, as sometimes happens, an infant has been born with massive and irreparable brain damage. The damage is so severe that the infant can never be any more than a "human vegetable," unable to talk, recognize other people, act independently of others, or develop a sense of self-awareness. The parents of the infant, realizing that they cannot hope for any improvement in their child's condition and being in any case unwilling to spend, or ask the state to spend, the thousands of dollars that would be needed annually for proper care of the infant, ask the doctor to kill the infant painlessly.

Should the doctor do what the parents ask? Legally, the doctor should not, and in this respect the law reflects the sanctity of life view. The life of every human being is sacred. Yet people who would say this about the infant do not object to the killing of nonhuman animals. How can they justify their different judgments? Adult chimpanzees, dogs, pigs, and members of many other species far surpass the brain-damaged infant in their ability to relate to others, act independently, be self-aware, and any other capacity that could reasonably be said to give value to life. With the most intensive care possible, some severely retarded infants can never achieve the intelligence level of a dog. Nor can we appeal to the concern of the infant's parents, since they themselves, in this imaginary example (and in some actual cases), do not want the infant kept alive. The only thing that distinguishes the infant from the animal, in the eyes of those who claim it has a "right to life," is that it is, biologically, a member of the species *Homo sapiens*, whereas chimpanzees, dogs, and pigs are not. But to use *this* difference as the basis for granting a right to life to the infant and not to the other animals is, of course, pure speciesism.[14] It is exactly the kind of arbitrary difference that the most crude and overt kind of racist uses in attempting to justify racial discrimination.

This does not mean that to avoid speciesism we must hold that it is as wrong to kill a dog as it is to kill a human being in full possession of his or her faculties. The only position that is irredeemably speciesist is the one that tries to make the boundary of the right to life run exactly parallel to the boundary of our own species. Those who hold the sanctity of life view do this, because while distinguishing sharply between human beings and other animals they allow no distinctions to be made within our own species, objecting to the killing of the severely retarded and the hopelessly senile as strongly as they object to the killing of normal adults.

To avoid speciesism we must allow that beings who are similar in all relevant respects have a similar right to life—and mere membership in our own biological species cannot be a morally relevant criterion for this right. Within these limits we could still hold, for instance, that it is worse to kill a normal adult human, with a capacity for self-awareness and the ability to plan for the future and have meaningful relations with others, than it is to kill a mouse, which presumably does not share all of these characteristics; or we might appeal to the close family and other personal ties that humans have but mice do not have to the same degree; or we might think that it is the consequences for other humans, who will be put in fear for their own lives, that makes the crucial difference; or we might think it is some combination of these factors, or other factors altogether.

Whatever criteria we choose, however, we will have to admit that they do not follow precisely the boundary of our own species. We may legitimately hold that there are some features of certain beings that make their lives more valuable than those of other beings; but there will surely be some nonhuman animals whose lives, by any standards, are more valuable than the lives of some humans. A chimpanzee, dog, or pig, for instance, will have a higher degree of self-awareness and a greater capacity for meaningful relations with others than a severely retarded infant or someone in a state of advanced senility. So if we base the right to life on these characteristics, we must grant these animals a right to life as good as, or better than, such retarded or senile humans.

This argument cuts both ways. It could be taken as showing that chimpanzees, dogs, and pigs, along with some other species, have a right to life and we commit a grave moral offense whenever we kill them, even when they are old and suffering and our intention is to put them out of their misery. Alternatively one could take the argument as showing that the severely retarded and hopelessly senile have no right to life and may be killed for quite trivial reasons, as we now kill animals.

Since the main concern of this book is with ethical questions having to do with animals and not with the morality of euthanasia, I shall not attempt to settle this issue finally.[15] I think it is reasonably clear, though, that while both of the positions just described avoid speciesism, neither is satisfactory. What we need is some middle position that would avoid speciesism but would not make the lives of the retarded and senile as cheap as the lives of pigs and dogs now are, or make the lives of pigs and dogs so sacrosanct that we think it wrong to put them out of hopeless misery. What we must do is bring nonhuman animals within our sphere of moral concern and cease to treat their lives as expendable for whatever trivial purposes we may have. At the same time, once we realize that the fact that a being is a member of our own species is not in itself enough to make it always wrong to kill that being, we may come to reconsider our policy of preserving human lives at all costs, even when there is no prospect of a meaningful life or of existence without terrible pain.

I conclude, then, that a rejection of speciesism does not imply that all lives are of equal worth. While self-awareness, the capacity to think ahead and have hopes and aspirations for the future, the capacity for meaningful relations with others, and so on are not relevant to the question of inflicting pain—since pain is pain, whatever other capacities, beyond the capacity to feel pain, the being may have—these capacities are relevant to the question of taking life. It is not arbitrary to hold that the life of a self-aware being, capable of abstract thought, of planning for the future, of complex acts of communication, and so on, is more valuable than the life of a being without these capacities. To see the difference between the issues of inflicting pain and taking life, consider how we would choose within our own species. If we had to choose to save the life of a normal human being or an intellectually disabled human being, we would probably choose to save the life of a normal human being; but if we had to choose between preventing pain in the normal human being or the intellectually disabled one—imagine that both have received painful but superficial injuries, and we have only enough painkiller for one of them—it is not nearly so clear how we ought to choose. The same is true when we consider other species. The evil of pain is, in itself, unaffected by the other characteristics of the being who feels the pain; the value of life is affected by these other characteristics. To give just one reason for this difference, to take the life of a being who has been hoping, planning, and working for some future goal is to deprive that being of the fulfillment of all those efforts; to take the life of a being with a mental capacity below the level needed to grasp that one is a being with a future—much less make plans for the future—cannot involve this particular kind of loss.[16]

Normally this will mean that if we have to choose between the life of a human being and the life of another animal, we should choose to save the life of the human; but there may be special cases in which the reverse holds true, because the human being in question does not have the capacities of a normal human being. So this view is not speciesist, although it may appear to be at first glance. The preference, in normal cases, for saving a human life over the life of an animal when a choice *has* to be made is a preference based on the characteristics that normal humans have, and not on the mere fact that they are members of our own species. This is why when we consider members of our own species who lack the characteristics of normal humans, we can no longer say that their lives are always to be preferred to those of other animals. This issue comes up in a practical way in the following chapter. In general, though, the question of when it is wrong to kill (painlessly) an animal is one to which we need give no precise answer. As long as we remember that we should give the same respect to the lives of animals as we give to the lives of those humans at a similar mental level, we shall not go far wrong.[17]

In any case, the conclusions that are argued for in this book flow from the principle of minimizing suffering alone. The idea that it is also wrong to kill animals painlessly gives some of these conclusions additional support that is welcome but strictly unnecessary. Interestingly enough, this is true even of the conclusion that we ought to become vegetarians, a conclusion that in the popular mind is generally based on some kind of absolute prohibition on killing.

Tools for Research

FROM *Animal Liberation*

AMONG THE TENS OF MILLIONS of experiments performed, only a few can possibly be regarded as contributing to important medical research. Huge numbers of animals are used in university departments such as forestry and psychology; many more are used for commercial purposes, to test new cosmetics, shampoos, food coloring agents, and other inessential items.[1] All this can happen only because of our prejudice against taking seriously the suffering of a being who is not a member of our own species. Typically, defenders of experiments on animals do not deny that animals suffer. They cannot deny the animals' suffering, because they need to stress the similarities between humans and other animals in order to claim that their experiments may have some relevance for human purposes. The experimenter who forces rats to choose between starvation and electric shock to see if they develop ulcers (which they do) does so because the rat has a nervous system very similar to a human being's, and presumably feels an electric shock in a similar way.

There has been opposition to experimenting on animals for a long time. This opposition has made little headway because experimenters, backed by commercial firms that profit by supplying laboratory animals and equipment, have been able to convince legislators and the public that opposition comes from uninformed fanatics who consider the interests of animals more important than the interests of human beings. But to be opposed to what is going on now it is not necessary to insist that all animal experiments stop immediately. All we need to say is that experiments serving no direct and urgent purpose should stop immediately, and in the remaining fields of research, we should, whenever possible, seek to replace experiments that involve animals with alternative methods that do not.

To understand why this seemingly modest change would be so important we need to know more about the experiments that are now being performed and have been performed for a century. Then we will be able to assess the claim by defenders of the present situation that experiments on animals are done only for important purposes. The following pages, therefore, describe some experiments on animals. Reading the reports of these experiments is not a pleasant experience; but we have an obligation to inform ourselves about what is done in our own community, especially since we are paying, through our taxes, for most of this research. If the animals have to undergo these experiments, the least we can do is read the reports and inform ourselves about them. That is why I have not attempted to tone down or gloss over some of the things that are done to animals. At the same time I have not tried to make these things worse than they really are. The reports that follow are all drawn from accounts written by the experimenters themselves and published by them in the scientific journals in which experimenters communicate with one another.

Such accounts are invariably more favorable to the experimenters than reports by an outside observer would be. There are two reasons for this. One is that the experimenters will not emphasize the suffering they have inflicted unless it is necessary to do so in order to communicate the results of the experiment, and this is rarely the case. Most suffering therefore goes unreported. Experimenters may consider it unnecessary to include in their reports any mention of what happens when electric shock devices are left on when they should have been turned off, when animals recover consciousness in the midst of an operation because of an improperly administered anesthetic, or when unattended animals sicken and die over the weekend. The second reason scientific journals are a source favorable to experimenters is that they include only those experiments that the experimenters and editors of the journals consider significant. A British government committee found that only about one quarter of experiments on animals ever found their way into print.[2] There is no reason to believe that accounts of a higher proportion of experiments are published in the United States; indeed, since the proportion of minor colleges with researchers of lesser talents is much higher in the United States than in Britain, it seems probable that an even smaller proportion of experiments yield results of any significance at all.

So in reading the following pages bear in mind that they are drawn from sources favorable to the experimenters; and if the results of the experiments do not appear to be of sufficient importance to justify the suffering they caused, remember that these examples are all taken from the small fraction of experiments that editors considered significant enough to publish. One last warning. The reports published in the journals always appear under the names of the experimenters. I have generally retained these names, since I see no reason to protect experimenters behind a cloak of anonymity. Nevertheless, it should not be assumed that the people named are especially evil or cruel people. They are doing what they were trained to do and what thousands of their colleagues do. The experiments are intended to illustrate not sadism on the part of individual experimenters but the institutionalized mentality of speciesism that makes it possible for these experimenters to do these things without serious consideration of the interests of the animals they are using.

CONSIDER EXPERIMENTS designed to produce what is known as "learned helplessness"—supposedly a model of depression in human beings. In 1953 R. Solomon, L. Kamin, and L. Wynne, experimenters at Harvard University, placed forty dogs in a device called a "shuttlebox," which consists of a box divided into two compartments, separated by a barrier. Initially the barrier was set at the height of the dog's back. Hundreds of intense electric shocks were delivered to the dog's feet through a grid floor. At first the dogs could escape the shock if they learned to jump the barrier into the other compartment. In an attempt to "discourage" one dog from jumping, the experimenters forced the dog to jump one hundred times onto a grid floor in the other compartment that also delivered a shock to the dog's feet. They said that as the dog jumped he gave a "sharp anticipatory yip which turned into a yelp when he landed on the electrified grid." They then blocked the passage between the compartments with a piece of plate glass and tested the dog again. The dog "jumped forward and smashed his head against the glass." The dogs began by showing symptoms such as "defecation, urination, yelping and shrieking, trembling, attacking the apparatus, and so on; but after ten or twelve days of trials dogs who were prevented from escaping shock ceased to resist. The experimenters reported themselves "impressed" by this, and concluded that a combination of the plate glass barrier and foot shock was "very effective" in eliminating jumping by dogs.[3]

This study showed that it was possible to induce a state of hopelessness and despair by repeated administration of severe inescapable shock. Such "learned helplessness" studies were further refined in the 1960s. One prominent experimenter was Martin Seligman of the University of Pennsylvania. He electrically shocked dogs through a steel grid floor with such intensity and persistence that the dogs stopped trying to escape and "learned" to be helpless. In one study, written with his colleagues Steven Maier and James Geer, Seligman describes his work as follows:

> When a normal, naive dog receives escape/avoidance training in a shuttlebox, the following behavior typically occurs: at the onset of electric shock the dog runs frantically about, defecating, urinating, and howling until it scrambles over the barrier and so escapes from shock. On the next trial the dog, running and howling, crosses the barrier more quickly, and so on, until efficient avoidance emerges.

Seligman altered this pattern by strapping dogs in harnesses and giving them shocks from which they had no means of escape. When the dogs were then placed in the original shuttlebox situation from which escape was possible, he found that

> such a dog reacts initially to shock in the shuttlebox in the same manner as the naive dog. However in dramatic contrast to the naive dog it soon stops running and remains silent until shock terminates. The dog does not cross the barrier and escape from shock. Rather it seems to "give up" and passively "accept" the shock. On succeeding trials the dog continues to fail to make escape movements and thus takes 50 seconds of severe, pulsating shock on each trial. . . . A dog previously exposed to inescapable shock . . . may take unlimited shock without escaping or avoiding at all.[4]

In the 1980s, psychologists have continued to carry out these "learned helplessness" experiments. At Temple University in Philadelphia, Philip Bersh and three other experimenters trained rats to recognize a warning light that alerted them to a shock that would be delivered within five seconds. Once they understood the warning, the rats could avoid the shock by moving into the safe compartment. After the rats had learned this avoidance behavior, the experimenters walled off the safe chamber and subjected them to prolonged periods of inescapable shock. Predictably, they found that even after escape was possible, the rats were unable to relearn the escape behavior quickly.[5]

Bersh and colleagues also subjected 372 rats to aversive shock testing to try to determine the relationship between Pavlovian conditioning and learned helplessness. They reported that the "implications of these findings for learned helplessness theory are not entirely clear" and that "a substantial number of questions remain."[6]

At the University of Tennessee at Martin, G. Brown, P. Smith, and R. Peters went to a lot of trouble to create a specially designed shuttlebox for goldfish, perhaps to see if Seligman's theory holds water. The experimenters subjected forty-five fish to sixty-five shock sessions each and concluded that "the data in the present study do not provide much support for Seligman's hypothesis that helplessness is learned."[7]

These experiments have inflicted acute, prolonged pain on many animals, first to prove a theory, then to disprove the theory, and finally to support modified versions of the original theory. Steven Maier, who with Seligman and Geer was a coauthor of the previously quoted report on inducing learned helplessness in dogs, has made a career out of perpetuating the learned helplessness model. Yet in a recent review article, Maier had this to say about the validity of this "animal model" of depression:

> It can be argued that there is not enough agreement about the characteristics, neurobiology, induction, and prevention/cure of depression to make such comparison meaningful. . . . It would thus appear unlikely that learned helplessness is a model of depression in any general sense.[8]

Although Maier tries to salvage something from this dismaying conclusion by saying that learned helplessness may constitute a model not of depression but of "stress and coping," he has effectively admitted that more than thirty years of animal experimentation have been a waste of time and of substantial amounts of taxpayers' money, quite apart from the immense amount of acute physical pain that they have caused.

WHEN ARE EXPERIMENTS on animals justifiable? Upon learning of the nature of many of the experiments carried out, some people react by saying that all experiments on animals should be prohibited immediately. But if we make our demands as absolute as this, the experimenters have a ready reply: Would we be prepared to let thousands of humans die if they could be saved by a single experiment on a single animal?

This question is, of course, purely hypothetical. There has never been and never could be a single experiment that saved thousands of lives. The way to reply to this hypothetical question is to pose another: Would the experimenters be prepared to carry out their experiment on a human orphan under six months old if that were the only way to save thousands of lives?

If the experimenters would not be prepared to use a human infant, then their readiness to use nonhuman animals reveals an unjustifiable form of discrimination on the basis of species, since adult apes, monkeys, dogs, cats, rats, and other animals are more aware of what is happening to them, more self-directing, and, so far as we can tell, at least as sensitive to pain as a human infant. (I have specified that the human infant be an orphan, to avoid the complications of the feelings of parents. Specifying the case in this way is, if anything, overgenerous to those defending the use of nonhuman animals in experiments, since mammals intended for experimental use are usually separated from their mothers at an early age, when the separation causes distress for both mother and young.)

So far as we know, human infants possess no morally relevant characteristic to a higher degree than adult nonhuman animals, unless we are to count the infants' potential as a characteristic that makes it wrong to experiment on them. Whether this characteristic should count is controversial—if we count it, we shall have to condemn abortion along with experiments on infants, since the potential of the infant and the fetus is the same. To avoid the complexities of this issue, however, we can alter our original question a little and assume that the infant is one with irreversible brain damage so severe as to rule out any mental development beyond the level of a six-month-old infant. There are, unfortunately, many such human beings, locked away in special wards throughout the country, some of them long since abandoned by their parents and other relatives, and, sadly, sometimes unloved by anyone else. Despite their mental deficiencies, the anatomy and physiology of these infants are in nearly all respects identical to those of normal humans. If, therefore, we were to force-feed them with large quantities of floor polish or drip concentrated solutions of cosmetics into their eyes, we would have a much more reliable indication of the safety of these products for humans than we now get by attempting to extrapolate the results of tests on a variety of other species. The LD50 tests, the Draize eye tests, the radiation experiments, the heatstroke experiments, and many others that cause suffering to nonhuman animals could have told us more about human reactions to the experimental situation if they had been carried out on severely brain-damaged humans instead of dogs or rabbits.

So whenever experimenters claim that their experiments are important enough to justify the use of animals, we should ask them whether they would be prepared to use a brain-damaged human being at a mental level similar to that of the animals they are planning to use. I cannot imagine that anyone would seriously propose carrying out the experiments described in this chapter on brain-damaged human beings. Occasionally it has become known that medical experiments have been performed on human beings without their consent; one case did concern institutionalized intellectually disabled children, who were given hepatitis.[9] When such harmful experiments on human beings become known, they usually lead to an outcry against the experimenters, and rightly so. They are, very often, a further example of the arrogance of the research worker who justifies everything on the grounds of increasing knowledge. But if the experimenter claims that the experiment is important enough to justify inflicting suffering on animals, why is it not important enough to justify inflicting suffering on humans at the same mental level? What difference is there between the two? Only that one is a member of our species and the other is not? But to appeal to that difference is to reveal a bias no more defensible than racism or any other form of arbitrary discrimination.

We have still not answered the question of when an experiment might be justifiable. It will not do to say "Never!" Putting morality in such black-and-white terms is appealing, because it eliminates the need to think about particular cases; but in extreme circumstances, such absolutist answers always break down. Torturing a human being is almost always wrong, but it is not absolutely wrong. If torture were the only way in which we could discover the location of a nuclear bomb hidden in a New York City basement and timed to go off within the hour, then torture would be justifiable. Similarly, if a single experiment could cure a disease like leukemia, that experiment would be justifiable. But in actual life the benefits are always more remote, and more often than not they are nonexistent. So how do we decide when an experiment is justifiable?

We have seen that experimenters reveal a bias in favor of their own species whenever they carry out experiments on nonhumans for purposes that they would not think justified using human beings, even brain-damaged ones. This principle gives us a guide toward an answer to our question. Since a speciesist bias, like a racist bias, is unjustifiable, an experiment cannot be justifiable unless the experiment is so important that the use of a brain-damaged human would also be justifiable.

This is not an absolutist principle. I do not believe that it could never be justifiable to experiment on a brain-damaged human. If it really were possible to save several lives by an experiment that would take just one life, and there were no other way those lives could be saved, it would be right to do the experiment. But this would be an extremely rare case. Certainly none of the experiments described in this chapter could pass this test. Admittedly, as with any dividing line, there would be a gray area where it was difficult to decide if an experiment could be justified. But we need not get distracted by such considerations now. As this chapter has shown, we are in the midst of an emergency in which appalling suffering is being inflicted on millions of animals for purposes that on any impartial view are obviously inadequate to justify the suffering. When we have ceased to carry out all those experiments, then there will be time enough to discuss what to do about the remaining ones which are claimed to be essential to save lives or prevent greater suffering.

THE DEFENDERS of animal experimentation are fond of telling us that animal experimentation has greatly increased our life expectancy. In the midst of the debate over reform of the British law on animal experimentation, for example, the Association of the British Pharmaceutical Industry ran a full-page advertisement in the *Guardian* under the headline "They say life begins at forty. Not so long ago, that's about when it ended." The advertisement went on to say that it is now considered to be a tragedy if a man dies in his forties, whereas in the nineteenth century it was commonplace to attend the funeral of a man in his forties, for the average life expectancy was only forty-two. The advertisement stated that "it is thanks largely to the breakthroughs that have been made through research which requires animals that most of us are able to live into our seventies."

Such claims are simply false. In fact, this particular advertisement was so blatantly misleading that a specialist in community medicine, Dr. David St. George, wrote to *The Lancet* saying, "The advertisement is good teaching material, since it illustrates two major errors in the interpretation of statistics." He also referred to Thomas McKeown's influential book *The Role of Medicine*,[10] which set off a debate about the relative contributions of social and environmental changes, as compared with medical intervention, in improvements in mortality since the mid-nineteenth century; and he added:

> This debate has been resolved, and it is now widely accepted that medical interventions had only a marginal effect on population mortality and mainly at a very late stage, after death rates had already fallen strikingly.[11]

J. B. and S. M. McKinley and R. Beaglehole reached a similar conclusion in a study of the decline of ten major infectious diseases in the United States. They showed that in every case except poliomyelitis the death rate had already fallen dramatically (presumably because of improved sanitation and diet) before any new form of medical treatment was introduced. Concentrating on the 40 percent fall in crude mortality in the United States between 1910 and 1984, they estimated "conservatively" that

> perhaps 3.5 percent of the fall in the overall death rate can be explained through medical interventions for the major infectious diseases. Indeed, given that it is precisely for these diseases that medicine claims most success in lowering mortality, 3.5 percent probably represents a reasonable upper-limit estimate of the total contribution of medical measures to the decline in infectious disease mortality in the United States.[12]

Remember that this 3.5 percent is a figure for all medical intervention. The contribution of animal experimentation itself can be, at most, only a fraction of this tiny contribution to the decline in mortality.

No doubt there are some fields of scientific research that will be hampered by any genuine consideration of the interests of animals used in experimentation. No doubt there have been some advances in knowledge which would not have been attained as easily without using animals. Examples of important discoveries often mentioned by those defending animal experimentation go back as far as Harvey's work on the circulation of blood. They include Banting and Best's discovery of insulin and its role in diabetes; the recognition of poliomyelitis as a virus and the development of a vaccine for it; several discoveries that served to make open heart surgery and coronary artery bypass graft surgery possible; and the understanding of our immune system and ways to overcome rejection of transplanted organs.[13] The claim that animal experimentation was essential in making these discoveries has been denied by some opponents of experimentation.[14] I do not intend to go into the controversy here. We have just seen that any knowledge gained from animal experimentation has made at best a very small contribution to our increased life span; its contribution to improving the quality of life is more difficult to estimate. In a more fundamental sense, the controversy over the benefits derived from animal experimentation is essentially unresolvable, because even if valuable discoveries were made using animals, we cannot say how successful medical research would have been if it had been compelled, from the outset, to develop alternative methods of investigation. Some discoveries would probably have been delayed, or perhaps not made at all; but many false leads would also not have been pursued, and it is possible that medicine would have developed in a very different and more efficacious direction, emphasizing healthy living rather than cures.

In any case, the ethical question of the justifiability of animal experimentation cannot be settled by pointing to its benefits for us, no matter how persuasive the evidence in favor of such benefits may be. The ethical principle of equal consideration of interests will rule out some means of obtaining knowledge. There is nothing sacred about the right to pursue knowledge. We already accept many restrictions on scientific enterprise. We do not believe that scientists have a general right to perform painful or lethal experiments on human beings without their consent, although there are many cases in which such experiments would advance knowledge far more rapidly than any other method. Now we need to broaden the scope of this existing restriction on scientific research.

Finally, it is important to realize that the major health problems of the world largely continue to exist, not because we do not know how to prevent disease and keep people healthy, but because no one is putting enough effort and money into doing what we already know how to do. The diseases that ravage Asia, Africa, Latin America, and the pockets of poverty in the industrialized West are diseases that, by and large, we know how to cure. They have been eliminated in communities that have adequate nutrition, sanitation, and health care. It has been estimated that 250,000 children die each week around the world, and that one quarter of these deaths are by dehydration caused by diarrhea. A simple treatment, already known and needing no animal experimentation, could prevent the deaths of these children.[15] Those who are genuinely concerned about improving health care would probably make a more effective contribution to human health if they left the laboratories and saw to it that our existing stock of medical knowledge reached those who need it most.

The exploitation of laboratory animals is part of the larger problem of speciesism and it is unlikely to be eliminated altogether until speciesism itself is eliminated. Surely one day, though, our children's children, reading about what was done in laboratories in the twentieth century, will feel the same sense of horror and incredulity at what otherwise civilized people could do that we now feel when we read about the atrocities of the Roman gladiatorial arenas or the eighteenth-century slave trade.

Down on the Factory Farm . . .

FROM *Animal Liberation*

or what happened to your dinner when it was still an animal

FOR MOST HUMAN BEINGS, especially those in modern urban and suburban communities, the most direct form of contact with nonhuman animals is at mealtime: we eat them. This simple fact is the key to our attitudes toward other animals, and also the key to what each one of us can do about changing these attitudes. The use and abuse of animals raised for food far exceeds, in sheer numbers of animals affected, any other kind of mistreatment. Over 100 million cows, pigs, and sheep are raised and slaughtered in the United States alone each year; and for poultry the figure is a staggering 5 billion. (That means that about eight thousand birds—mostly chickens—will have been slaughtered in the time it takes you to read this page.) It is here, on our dinner table and in our neighborhood supermarket or butcher's shop, that we are brought into direct touch with the most extensive exploitation of other species that has ever existed.

In general, we are ignorant of the abuse of living creatures that lies behind the food we eat. Buying food in a store or restaurant is the culmination of a long process, of which all but the end product is delicately screened from our eyes. We buy our meat and poultry in neat plastic packages. It hardly bleeds. There is no reason to associate this package with a living, breathing, walking, suffering animal. The very words we use conceal its origins, we eat beef, not bull, steer, or cow; and pork, not pig—

although for some reason we seem to find it easier to face the true nature of a leg of lamb. The term "meat" is itself deceptive. It originally meant any solid food, not necessarily the flesh of animals. This image still lingers in an expression like "nut meat," which seems to imply a substitute for "flesh meat" but actually has an equally good claim to be called "meat" in its own right. By using the more general "meat" we avoid facing the fact that what we are eating is really flesh.

These verbal disguises are merely the top layer of a much deeper ignorance of the origin of our food. Consider the images conjured up by the word "farm": a house; a barn; a flock of hens, overseen by a strutting rooster, scratching around the farmyard; a herd of cows being brought in from the fields for milking; and perhaps a sow rooting around in the orchard with a litter of squealing piglets running excitedly behind her.

Very few farms were ever as idyllic as that traditional image would have us believe. Yet we still think of a farm as a pleasant place, far removed from our own industrial, profit-conscious city life. Of those few who think about the lives of animals on farms, not many know much about modern methods of raising animals. Some people wonder whether animals are slaughtered painlessly, and anyone who has followed a truckload of cattle on the road will probably know that farm animals are transported in extremely crowded conditions; but not many suspect that transportation and slaughter are anything more than the brief and inevitable conclusion of a life of ease and contentment, a life that contains the natural pleasures of animal existence without the hardships that wild animals must endure in their struggle for survival.

These comfortable assumptions bear little relation to the realities of modern farming. For a start, farming is no longer controlled by simple country folk. During the last fifty years, large corporations and assembly-line methods of production have turned agriculture into agribusiness. The process began when big companies gained control of poultry production, once the preserve of the farmer's wife. Today, fifty large corporations virtually control all poultry production in the United States. In the field of egg production, where fifty years ago a big producer might have had three thousand laying hens, today many producers have more than 500,000 layers, and the largest have over 10 million. The remaining small producers have had to adopt the methods of the giants or else go out of business. Companies that had no connection with agriculture have become farmers on a huge scale in order to gain tax concessions or to diversify profits. Greyhound Corporation now produces turkeys, and your roast beef may have come from John Hancock Mutual Life Insurance or from one of a dozen oil companies that have invested in cattle feeding, building feedlots that hold 100,000 or more cattle.[1]

The big corporations and those who must compete with them are not concerned with a sense of harmony among plants, animals, and nature. Farming is competitive, and the methods adopted are those that cut costs and increase production. So farming is now "factory farming." Animals are treated like machines that convert low-priced fodder into high-priced flesh, and any innovation will be used if it results in a cheaper "conversion ratio." Most of this chapter is simply a description of these methods, and of what they mean for the animals to whom they are applied. The aim is to demonstrate that under these methods animals lead miserable lives from birth to slaughter. Once again, however, my point is not that the people who do these things to the animals are cruel and wicked. On the contrary, the attitudes of the consumers and the producers are not fundamentally different. The farming methods I am about to describe are merely the logical application of the attitudes and prejudices that are discussed elsewhere in this book. Once we place nonhuman animals outside our sphere of moral consideration and treat them as things we use to satisfy our own desires, the outcome is predictable.

OF ALL THE FORMS of intensive farming now practiced, the veal industry ranks as the most morally repugnant. The essence of veal raising is the feeding of a high-protein food to confined, anemic calves in a manner that will produce a tender, pale-colored flesh that will be served to the patrons of expensive restaurants. Fortunately this industry does not compare in size with poultry, beef, or pig production; nevertheless it is worth our attention because it represents an extreme, both in the degree of exploitation to which it subjects the animals and in its absurd inefficiency as a method of providing people with nourishment.

Veal is the flesh of a young calf. The term was originally reserved for calves killed before they had been weaned from their mothers. The flesh of these very young animals was paler and more tender than that of a calf who had begun to eat grass; but there was not much of it, since calves begin to eat grass when they are a few weeks old and still very small. The small amount available came from the unwanted male calves produced by the dairy industry. A day or two after being born they were trucked to market where, hungry and frightened by the strange surroundings and the absence of their mothers, they were sold for immediate delivery to the slaughterhouse.

Then in the 1950s veal producers in Holland found a way to keep the calf alive longer without the flesh becoming red or less tender. The trick depends on keeping the calf in highly unnatural conditions. If calves were left to grow up outside they would romp around the fields, developing muscles that would toughen their flesh and burning up calories that the producer must replace with costly feed. At the same time they would eat grass, and their flesh would lose the pale color that the flesh of newborn calves has. So the specialist veal producers take their calves straight from the auction ring to a confinement unit. Here, in a converted barn or specially built shed, they have rows of wooden stalls, each 1 foot 10 inches wide by 4 feet 6 inches long. It has a slatted wooden floor, raised above the concrete floor of the shed. The calves are tethered by a chain around the neck to prevent them from turning in their stalls when they are small. (The chain may be removed when the calves grow too big to turn around in such narrow stalls.) The stall has no straw or other bedding, since the calves might eat it, spoiling the paleness of their flesh. They leave their stalls only to be taken out to slaughter. They are fed a totally liquid diet, based on nonfat milk powder with vitamins, minerals, and growth-promoting drugs added. Thus the calves live for the next sixteen weeks. The beauty of the system, from the producers' point of view, is that at this age the veal calf may weigh as much as four hundred pounds, instead of the ninety-odd pounds that newborn calves weigh; and since veal fetches a premium price, rearing veal calves in this manner is a profitable occupation.

This method of raising calves was introduced to the United States in 1962 by Provimi, Inc., a feed manufacturer based in Watertown, Wisconsin. Its name comes from the "proteins, vitamins, and minerals" of which its feeds are composed—ingredients that, one might think, could be put to better use than raising veal. Provimi, according to its own boast, created this "new and complete concept in veal raising" and it is still by far the largest company in the business, controlling 50 to 75 percent of the domestic market. Its interest in promoting veal production lies in developing a market for its feed. Describing what it considered "optimum veal production," Provimi's now defunct newssheet, *The Stall Street Journal*, gives us an insight into the nature of the industry, which in the United States and some European countries has remained essentially unchanged since its introduction:

> The dual aims of veal production are firstly, to produce a calf of the greatest weight in the shortest possible time and secondly, to keep its meat as light colored as possible to fulfill the consumer's requirement. All at a profit commensurate to the risk and investment involved.[2]

The narrow stalls and their slatted wooden floors are a serious source of discomfort to the calves. When the calves grow larger, they cannot even stand up and lie down without difficulty. As a report from a research group headed by Professor John Webster of the animal husbandry unit at the School of Veterinary Science, University of Bristol, in England, noted:

> Veal calves in crates 750 mm wide cannot, of course, lie flat with their legs extended. . . . Calves may lie like this when they feel warm and wish to lose heat. . . . Well-grown veal calves at air temperatures above 20 degrees C [68 degrees F] may be uncomfortably hot. Denying them the opportunity to adopt a position designed to maximise heat loss only makes things worse. . . . Veal calves in boxes over the age of 10 weeks were unable to adopt a normal sleeping position with their heads tucked into their sides. We conclude that denying veal calves the opportunity to adopt a normal sleeping posture is a significant insult to welfare. To overcome this, the crates would need to be at least 900 mm wide.[3]

American readers should note that 750 millimeters is equivalent to 2 feet 6 inches, and 900 millimeters to 3 feet, both considerably more than the standard 1-foot 10-inch crates used in the United States.

The crates are also too narrow to permit the calf to turn around. This is another source of frustration. In addition, a stall too narrow to turn around in is also too narrow to groom comfortably in; and calves have an innate desire to twist their heads around and groom themselves with their tongues. As the University of Bristol researchers said:

> Because veal calves grow so fast and produce so much heat they tend to shed their coats at about 10 weeks of age. During this time they have a great urge to groom themselves. They are also particularly prone to infestation with external parasites, especially in mild, humid conditions. Veal calves in crates cannot reach much of their body. We conclude that denying the veal calf the opportunity to groom itself thoroughly is an unacceptable insult to welfare whether this is achieved by constraining its freedom of movement or, worse, by the use of a muzzle.[4]

A slatted wooden floor without any bedding is hard and uncomfortable; it is rough on the calves' knees as they get up and lie down. In addition, animals with hooves are uncomfortable on slatted floors. A slatted floor is like a cattle grid, which cattle always avoid, except that the slats are closer together. The spaces, however, must still be large enough to allow most of the manure to fall or be washed through, and this means that they are large enough to make the calves uncomfortable on them. The Bristol team described the young calves as "for some days insecure and reluctant to change position."

The young calves sorely miss their mothers. They also miss something to suck on. The urge to suck is strong in a baby calf, as it is in a baby human. These calves have no teat to suck on, nor do they have any substitute. From their first day in confinement—which may well be only the third or fourth day of their lives—they drink from a plastic bucket. Attempts have been made to feed calves through artificial teats, but the task of keeping the teats clean and sterile is apparently not worth the producer's trouble. It is common to see calves frantically trying to suck some part of their stalls, although there is usually nothing suitable; and if you offer a veal calf your finger you will find that he immediately begins to suck on it, as human babies suck their thumbs.

Later the calf develops a need to ruminate—that is, to take in roughage and chew the cud. But roughage is strictly forbidden because it contains iron and will darken the flesh, so, again, the calf may resort to vain attempts to chew the sides of his stall. Digestive disorders, including stomach ulcers, are common in veal calves. So is chronic diarrhea. To quote the Bristol study once again:

> The calves are deprived of dry feed. This completely distorts the normal development of the rumen and encourages the development of hair balls which may also lead to chronic indigestion.[5]

As if this were not enough, the calf is deliberately kept anemic. Provimi's *Stall Street Journal* explains why:

> Color of veal is one of the primary factors involved in obtaining "top-dollar" returns from the fancy veal markets. . . . "Light color" veal is a premium item much in demand at better clubs, hotels and restaurants. "Light color" or pink veal is partly associated with the amount of iron in the muscle of the calves.[6]

So Provimi's feeds, like those of other manufacturers of veal feeds, are deliberately kept low in iron. A normal calf would obtain iron from grass and other forms of roughage, but since veal calves are not allowed this, they become anemic. Pale pink flesh is in fact anemic flesh. The demand for flesh of this color is a matter of snob appeal. The color does not affect the taste and it certainly does not make the flesh more nourishing—it just means that it lacks iron.

The anemia is, of course, controlled. Without any iron at all the calves would drop dead. With a normal intake their flesh will not fetch as much per pound. So a balance is struck which keeps the flesh pale and the calves—or most of them—on their feet long enough for them to reach market weight. The calves, however, are unhealthy and anemic animals. Kept deliberately short of iron, they develop a craving for it and will lick any iron fittings in their stalls. This explains the use of wooden stalls. As Provimi tells its customers:

> The main reason for using hardwood instead of metal boxstalls is that metal may affect the light veal color. . . . Keep all iron out of reach of your calves.[7]

And again:

> It is also necessary that calves do not have access to a continuous source of iron. (Water supplied should be checked. If a high level of iron [excess of 0.5 ppm] is present an iron filter should be considered.) Calf crates should be constructed so calves have no access to rusty metal.[8]

The anemic calf's insatiable craving for iron is one of the reasons the producer is anxious to prevent him from turning around in his stall. Although calves, like pigs, normally prefer not to go near their own urine or manure, urine does contain some iron. The desire for iron is strong enough to overcome the natural repugnance, and the anemic calves will lick the slats that are saturated with urine. The producer does not like this, because it gives calves a little iron and because in licking the slats the calves may pick up infections from their manure, which falls on the same spot as their urine.

We have seen that in the view of Provimi, Inc., the twin aims of veal production are producing a calf of the greatest possible weight in the shortest possible time and keeping the meat as light in color as possible.

We have seen what is done to achieve the second of these aims, but there is more to be said about the techniques used to achieve fast growth.

To make animals grow quickly they must take in as much food as possible, and they must use up as little of this food as possible in their daily life. To see that the veal calf takes in as much as possible, most calves are given no water. Their only source of liquid is their food—the rich milk replacer based on powdered milk and added fat. Since the buildings in which they are housed are kept warm, the thirsty animals take in more of their food than they would do if they could drink water. A common result of this overeating is that the calves break out in a sweat, rather like, it has been said, an executive who has had too much to eat too quickly.[9] In sweating, the calf loses moisture, which makes him thirsty, so that he overeats again next time. By most standards this process is an unhealthy one, but by the standards of the veal producer aiming at producing the heaviest calf in the shortest possible time, the long-term health of the animal is irrelevant, so long as he survives to be taken to market; and so Provimi advises that sweating is a sign that "the calf is healthy and growing at capacity."[10]

Getting the calf to overeat is half the battle; the other half is ensuring that as much as possible of what has been eaten goes toward putting on weight. Confining the calf so that he cannot exercise is one requirement for achieving this aim. Keeping the barn warm also contributes to it, since a cold calf burns calories just to keep warm. Even warm calves in their stalls are apt to become restless, however, for they have nothing to do all day except at their two mealtimes. A Dutch researcher has written:

> Veal calves suffer from the inability to do something. . . . The food-intake of a veal calf takes only 20 minutes a day! Besides that there is nothing the animal can do. . . . One can observe teeth grinding, tail wagging, tongue swaying and other stereotype behavior. . . . Such stereotype movements can be regarded as a reaction to a lack of occupation.[11]

To reduce the restlessness of their bored calves, many veal producers leave the animals in the dark at all times, except when they are being fed. Since the veal sheds are normally windowless, this simply means turning off the lights. Thus the calves, already missing most of the affection, activity and stimulation that their natures require, are deprived of visual stimulation and of contact with other calves for more than twenty-two hours out of every twenty-four. Illnesses have been found to be more persistent in dark sheds.[12]

Calves kept in this manner are unhappy and unhealthy animals. Despite the fact that the veal producer selects only the strongest, healthiest calves to begin with, uses a medicated feed as a routine measure, and gives additional injections at the slightest sign of illness, digestive, respiratory, and infectious diseases are widespread. It is common for a veal producer to find that one in ten of a batch of calves does not survive the fifteen weeks of confinement. Between 10 and 15 percent mortality over such a short period would be disastrous for anyone raising calves for beef, but veal producers can tolerate this loss because the high-priced restaurants are prepared to pay well for their products.

Given the cozy relationship that normally exists between veterinarians working with farm animals and intensive producers (it is, after all, the owners, not the animals, who pay the bills), it gives us some indication of the extreme conditions under which veal calves are kept to learn that this is one aspect of animal production that has strained relations between veterinarians and producers. A 1982 issue of *The Vealer* reports:

> Besides waiting too long to call veterinarians for a really sick calf, vets do not look favorable [sic] on relations with veal growers because they have long defied accepted agricultural methods. The feeding of long hay to livestock, in order to maintain a proper digestive system, has been considered a sound practice for years.[13]

The one bright spot in this sorry tale is that the conditions created by the veal crates are so appalling for animal welfare that British government regulations now require that a calf must be able to turn around without difficulty, must be fed a daily diet containing "sufficient iron to maintain it in full health and vigour," and must receive enough fiber to allow normal development of the rumen.[14] These are minimal welfare requirements and still fall well short of satisfying the needs of calves; but they are violated by almost all the veal units in the United States and by many in Europe.

If the reader will recall that this whole laborious, wasteful, and painful process of veal raising exists for the sole purpose of pandering to people who insist on pale, soft veal, no further comment should be needed.

A Vegetarian Philosophy

FROM *Consuming Passions*

ISSUES REGARDING EATING meat were highlighted in 1997 by the longest trial in British legal history. *McDonald's Corporation and McDonald's Restaurants Limited* v. *Steel and Morris,* better known as the "McLibel" trial, ran for 313 days and heard 180 witnesses. In suing Helen Steel and David Morris, two activists involved with the London Greenpeace organization, McDonald's put on trial the way in which its fast-food products are produced, packaged, advertised, and sold, as well as their nutritional value, the environmental impact of producing them, and the treatment of the animals whose flesh and eggs are made into that food. . . .

The case provided a remarkable opportunity for weighing up evidence for and against modern agribusiness methods. The leaflet "What's Wrong with McDonald's" that provoked the defamation suit had a row of McDonald's arches along the top of each page. Two of these arches bore the words "McMurder" and "McTorture." One section below was headed "In what way are McDonald's responsible for torture and murder?" The leaflet answered the question as follows:

> The menu at McDonald's is based on meat. They sell millions of burgers every day in 35 countries throughout the world. This means the constant slaughter, day by day, of animals born and bred solely to be turned

> into McDonald's products. Some of them—especially chickens and pigs—spend their lives in the entirely artificial conditions of huge factory farms, with no access to air or sunshine and no freedom of movement. Their deaths are bloody and barbaric.

McDonald's claimed that the leaflet meant that the company was responsible for the inhumane torture and murder of cattle, chicken, and pigs, and that this was defamatory. In considering this claim, Mr. Justice Bell based his judgment on what he took to be attitudes that were generally accepted in Britain. Thus for the epithet "McTorture" to be justified, he held, it would not be enough for Steel and Morris to show that animals were under stress or suffered some pain or discomfort:

> Merely containing, handling and transporting an animal may cause it stress; and taking it to slaughter certainly may do so. But I do not believe that the ordinary reasonable person believes any of these things to be cruel, provided that the necessary stress, or discomfort or even pain is kept to a reasonably acceptable level. That ordinary person may know little about the detail of farming and slaughtering methods but he must find a certain amount of stress, discomfort or even pain acceptable and not to be criticised as cruel.

By the end of the trial, however, Mr. Justice Bell found that the stress, discomfort, and pain inflicted on some animals amounted to more than this acceptable level, and hence did constitute a "cruel practice" for which McDonald's was "culpably responsible." Chickens, laying hens, and sows, he said, kept in individual stalls suffered from "severe restriction of movement" which "is cruel." He also found a number of other cruel practices in the production of chickens, including the restricted diet fed to breeding birds, which leaves them permanently hungry; the injuries inflicted on chickens by catchers stuffing 600 birds an hour into crates to take them to slaughter; and the failure of the stunning apparatus to ensure that all birds are stunned before they have their throats cut. Judging by entirely conventional moral standards, Mr. Justice Bell held these practices to be cruel, and McDonald's to be culpably responsible for them.

It was not libelous to describe McDonald's as "McTorture," because the charge was substantially true. What follows from this judgment about the morality of buying and eating intensively raised chickens, pig products that come from the offspring of sows kept in stalls, or eggs laid by hens kept in battery cages? Surely that, too, must be wrong?

This claim has been challenged. At a conference dinner some years ago I found myself sitting opposite a Buddhist philosopher from Thailand. As we helped ourselves to the lavish buffet, I avoided the various forms of meat being offered, but the Thai philosopher did not. When I asked him how he reconciled the dinner he had chosen with the first precept of Buddhism, which tells us to avoid harming sentient beings, he told me that in the Buddhist tradition it is wrong to eat meat only if you have reason to believe that the animal was killed specially for you. The meat he had taken, however, was not from animals killed specially for him; the animals would have died anyway, even if he were a strict vegetarian or had not been in that city at all. Hence, by eating it, he was not harming any animals.

I was unable to convince my dinner companion that this defense of meat eating was better suited to a time when a peasant family might kill an animal especially to have something to put in the begging bowl of a wandering monk than it is to our own era. The flaw in the defense is the disregard of the link between the meat I eat today and the future killing of animals. Granted, the chicken lying in the supermarket freezer today would have died even if I had never existed; but the fact that I take the chicken from the freezer, and ignore the tofu on a nearby shelf, has something to do with the number of chickens, or blocks of tofu, the supermarket will order next week and thus contributes, in a small way, to the future growth or decline of the chicken and tofu industries. That is what the laws of supply and demand are all about.

Some defenders of a variant of the ancient Buddhist line may still want to argue that one chicken fewer sold makes no perceptible difference to the chicken producers, and therefore there can be nothing wrong with buying chicken. The division of moral responsibility in a situation of this kind does raise some interesting issues, but it is a fallacy to argue that a person can do wrong only by making a perceptible harm. The Oxford philosopher Jonathan Glover has explored the implications of this refusal to accept the divisibility of responsibility in an entertaining article called "It makes no difference whether or not I do it" (*Proceedings of the Aristotelian Society*, 1975).

Glover imagines that in a village, 100 people are about to eat lunch. Each has a bowl containing 100 beans. Suddenly, 100 hungry bandits swoop down on the village. Each bandit takes the contents of the bowl of one villager, eats it, and gallops off. Next week, the bandits plan to do it again, but one of their number is afflicted by doubts about whether it is right to steal from the poor. These doubts are set to rest by another of their number who proposes that each bandit, instead of eating the entire contents of the bowl of one villager, should take one bean from every villager's bowl. Since the loss of one bean cannot make a perceptible difference to any villager, no bandit will have harmed anyone. The bandits follow this plan, each taking a solitary bean from 100 bowls. The villagers are just as hungry as they were the previous week, but the bandits can all sleep well on their full stomachs, knowing that none of them has harmed anyone.

Glover's example shows the absurdity of denying that we are each responsible for a share of the harms we collectively cause, even if each of us makes no perceptible difference. McDonald's has a far bigger impact on the practices of the chicken, egg, and pig industries than any individual consumer; but McDonald's itself would be powerless if no one ate at its restaurants. Collectively, all consumers of animal products are responsible for the existence of the cruel practices involved in producing them. In the absence of special circumstances, a portion of this responsibility must be attributed to each purchaser.

Without in any way departing from a conventional moral attitude toward animals, then, we have reached the conclusion that eating intensively produced chicken, battery eggs, and some pig products is wrong. This is, of course, well short of an argument for vegetarianism. Mr. Justice Bell found "cruel practices" only in these areas of McDonald's food production. But he did not find that McDonald's beef is "cruelty-free." He did not consider that question, because he drew a distinction between McDonald's responsibility for practices in the beef and dairy industries and those in the chicken, egg, and pig industries. McDonald's chickens, eggs, and pig products are supplied by a relatively small number of very large producers, over whose practices the corporation could quite easily have a major influence. On the other hand, McDonald's beef and dairy requirements came from a very large number of producers; and in respect of whose methods, Mr. Justice Bell held, "there was no evidence from which I could infer that [McDonald's] would have any effective influence, should it try to exert it." Whatever one may think of that view—it seems highly implausible to me—the judge, in accepting it, decided not to address the evidence presented to him of cruelty in the raising of cattle, so that no conclusions either way can be drawn.

This does not mean that the trial itself had nothing to say about animal suffering in general. McDonald's called as a witness Mr. David

Walker, chief executive of one of McDonald's major United Kingdom suppliers, McKey Food Services Ltd. In cross-examination, Helen Steel asked Walker whether it was true that, "as the result of the meat industry, the suffering of animals is inevitable." Walker replied: "The answer to that must be 'yes.' "

Walker's admission raises a serious question about the ethics of the meat industry: how much suffering are we justified in inflicting on animals in order to turn them into meat, or to use their eggs or milk?

The case for vegetarianism is at its strongest when we see it as a moral protest against our use of animals as mere things, to be exploited for our convenience in whatever way makes them most cheaply available to us. Only the tiniest fraction of the tens of billions of farm animals slaughtered for food each year—the figure for the United States alone is nine billion—were treated during their lives in ways that respected their interests. Questions about the wrongness of killing in itself are not relevant to the moral issue of eating meat or eggs from factory-farmed animals, as most people in developed countries do. Even when animals are roaming freely over large areas, as sheep and cattle do in Australia, operations like hot-iron branding, castration, and dehorning are carried out without any regard for the animals' capacity to suffer. The same is true of handling and transport prior to slaughter. In the light of these facts, the issue to focus on is not whether there are some circumstances in which it could be right to eat meat, but on what we can do to avoid contributing to this immense amount of animal suffering.

The answer is to boycott all meat and eggs produced by large-scale commercial methods of animal production, and encourage others to do the same. Consideration for the interests of animals alone is enough justification for this response, but the case is further strengthened by the environmental problems that the meat industry causes. Although Mr. Justice Bell found that the allegations directed at McDonald's regarding its contribution to the destruction of rain forests were not true, the meat industry as a whole can take little comfort from that, because Bell accepted evidence that cattle-ranching, particularly in Brazil, had contributed to the clearing of vast areas of rain forest. The problem for David Morris and Helen Steel was that they did not convince the judge that the meat used by McDonald's came from these regions. So the meat industry as a whole remains culpable for the loss of rain forest and for all the consequences of that, from global warming to the deaths of indigenous people fighting to defend their way of life.

Environmentalists are increasingly recognizing that the choice of what we eat is an environmental issue. Animals raised in sheds or on feedlots eat grains or soybeans, and they use most of the food value of these products simply in order to maintain basic functions and develop unpalatable parts of the body like bones and skin. To convert eight or nine kilos of grain protein into a single kilo of animal protein wastes land, energy, and water. On a crowded planet with a growing human population, that is a luxury that we are becoming increasingly unable to afford.

Intensive animal production is a heavy user of fossil fuels and a major source of pollution of both air and water. It releases large quantities of methane and other greenhouse gases into the atmosphere. We are risking unpredictable changes to the climate of our planet—which means, ultimately, the lives of billions of people, not to mention the extinction of untold thousands of species of plants and animals unable to cope with changing conditions—for the sake of more hamburgers. A diet heavy in animal products, catered to by intensive animal production, is a disaster for animals, the environment, and the health of those who eat it.

A Recipe

THIS RECIPE IS VEGAN, very simple, nutritious, and tasty. It's also eaten by hundreds of millions of people every day.

DAL

2 tablespoons oil
1 onion, chopped
2 cloves garlic, crushed
1 cup dry red lentils
3 cups water
bay leaf
1 cinnamon stick
1 teaspoon medium curry powder or to taste
1 14-ounce can of chopped tomatoes or equivalent chopped fresh tomatoes
2 ounces creamed coconut or half cup coconut milk (optional)
Juice of 1 lemon (optional)
Salt to taste

In a deep frying pan, heat the oil and fry the onion and garlic until translucent. Add the lentils and fry them for a minute or two, then add the water, bay leaf, cinnamon stick, and curry powder. Stir, bring to a boil, then let simmer for twenty minutes, adding a little more water from time to time if it gets dry. Add the tomatoes and simmer another ten minutes. By now the lentils should be very soft. Add the creamed coconut or coconut milk and lemon juice, if using, and salt to taste. Remove cinnamon stick and bay leaf before serving.

The final product should flow freely—add more water if it is too thick. It is usually served over rice, with some lime pickle and mango chutney. Sliced banana is another good accompaniment, and so too are pappadams.

Bridging the Gap

An Unusual Institution

In the Netherlands a few years ago, an observer reported on the lives of some people confined in a new kind of institution. These people had a special condition that did not handicap them at all physically, but intellectually they were well below the normal human level; they could not speak, although they made noises and gestures. In the institutions in which such people were usually kept, they tended to spend much of their time making repetitive movements and rocking their bodies to and fro. This institution was an unusual one, in that its policy was to allow the inmates the maximum possible freedom to live their own lives and form their own community. This freedom extended even to sexual relationships, which led to pregnancy, birth, and child-rearing.

Under these circumstances the behavior of the inmates was far more varied than in the more conventional institutional settings. They rarely spent time alone, and they appeared to have no difficulty in understanding each other's gestures and vocalizations. They were physically active, spending a lot of time outside, where they had access to about two acres of relatively natural forest surrounded by a wall. They cooperated in many activities, including, on one occasion—to the consternation of the supervisors—an attempt to escape that involved carrying a large fallen branch to one of the walls and propping it up as a kind of ladder, making it possible to climb over the wall.

The observer was particularly interested in what he called the "politics" of the community. A defined leader soon emerged. His leadership—and it was always a "he"—depended, however, on the support of other members of the group. The leader had privileges, but also, it seemed, obligations. He had to cultivate the favor of others by sharing food and other treats. Fights would develop from time to time, but they would usually be followed by some conciliatory gestures, so that the loser could be readmitted into the society of the leader. If the leader became isolated and allowed the others to form a coalition against him, his days as leader were numbered.

A simple ethical code could also be detected within the community. Its two basic rules, the observer commented, could be summed up as "One good turn deserves another," and "An eye for an eye and a tooth for a tooth." The breach of the first of these rules apparently led to a sense of being wronged. For example, on one occasion Henk was fighting with Jan, and Gert came to Jan's assistance. Later, Henk attacked Gert, who gestured to Jan for assistance, but Jan did nothing. After the fight between Gert and Henk was over, Gert furiously attacked Jan.

The mothers were, with one exception, competent at nursing and rearing their children. The mother-child relationships were close and lasted many years. The death of a baby led to prolonged grieving behavior. Because sexual relationships were not monogamous, it was not always possible to tell who the father of the child was, and fathers did not play a significant role in the rearing of the children.

In view of the very limited mental capacities that these inmates had been considered to possess, the observer was impressed by instances of behavior that clearly showed intelligent planning. In one example, two young mothers were having difficulty in stopping their small children from fighting. An older mother, a considerable authority figure in the community, was dozing nearby. One of the younger mothers woke her and pointed to the squabbling children. The older mother made the appropriate noises and gestures, and the children, suitably intimidated, stopped fighting. The older mother then went back to her nap.

In order to see just how far ahead these people could think, the observer devised an ingenious test of problem-solving ability. One inmate was presented with two series of five locked clear plastic boxes, each of which opened with a different, but readily identifiable, key. The keys were visible in the boxes. One series of five boxes led to a food treat, whereas the other series led to an empty box. The key to the first box in each series lay beside it. It was necessary to begin by choosing one of these two initial boxes; to succeed, one had to work mentally through the five boxes to see which initial choice would lead one to the box with the treat. The inmate was able to succeed in this complex task.

The inmates' own awareness of what they were doing was well shown by their extensive practice of deceit. On one occasion, after a fight, it was noticeable that the loser limped badly when in the presence of the victor, but not when alone; presumably by pretending to be more seriously hurt than he really was, he hoped for some kind of sympathy, or at least mercy, from his conqueror.

But the most elaborate forms of deceit were concerned with—no surprise here for any observer of human behavior—sexual relationships. Although, as already mentioned, sexual relationships were not monogamous, the leader tried to prevent others from having sexual relationships with his favorites. To get around this, flirtations leading up to sexual intercourse were conducted with a good deal of discretion, so as not to attract the leader's attention.

I have described this community in some detail because I want to raise an ethical question about the way in which people with this condition were regarded by those who looked after them. In the eyes of their supervisors the inmates did not have the same kind of right to life as normal human beings. Though treated with care and consideration for their welfare, they were seen as clearly inferior, and their lives were accorded much less value than the lives of normal human beings. When one of them was killed, in the course of a dispute over who should be leader, the killing was not considered equivalent to the killing of a normal human being. Moreover, in other institutions—not the one I have just described, but different ones, both in the Netherlands and overseas—people with this condition are deliberately infected with diseases such as hepatitis or AIDS, in order to test the efficacy of experimental drugs or vaccines. In some cases they die as a result of the experiment.

How should we regard this situation? Is it a moral outrage? Or is it right and proper, given the more limited intellectual capacities of these people?

Your answer to this question will almost certainly vary according to the mental image you formed of the inmates of the community I have described. I referred to them as "people." In doing so I had in mind the definition of "person" offered by the seventeenth-century British philosopher John Locke: "A thinking intelligent being that has reason and reflection and can consider itself as itself, the same thinking thing, in different times and places."[1] But because the term "person," like "people," is commonly used only of members of the species *Homo sapiens,* my use of the term may have led you to think that the community I was describing was a community of intellectually disabled human beings. My use of Dutch names probably reinforced that assumption. In that case, you probably also thought it very wrong that the lives of these people were accorded less value than those of normal human beings, and the mention of their use as experimental subjects very likely caused shock and a sense of outrage.

Perhaps many of you, however, were able to guess that the description was not one of human beings at all. The "special condition" that these people have is their membership of the species *Pan troglodytes.* They are a community of chimpanzees, living in Arnhem Zoo.[2] If you guessed this, you may not have been so shocked by what I told you about the way in which the supervisors thought of the value of life of the inmates as being markedly less than that of normal humans—perhaps not even by the use of the inmates in lethal experiments.

The distinction between human beings and all other animals is fundamental to our ethical attitudes toward ourselves, toward the rest of nature, and toward ethical problems of life and death. Here is an example of the way in which this distinction is taken to absurd lengths.

Whose Organs May We Take?

BABY VALENTINA WAS BORN in Palermo, Italy, in April 1992. She was an anencephalic—that is, she was born with all of her brain, except the brain stem, missing. This meant that she would never be able to be conscious, to respond to her mother's smile, or to experience anything at all. Such babies usually die within a few days of birth. Valentina's parents, seeking to salvage something out of a birth that was so much less than they had expected, offered her as an organ donor. Amidst heated public debate, the Italian court ruled that this could not be permitted. To take the heart or any other vital organ from a living human being, even one with nothing more than a brain stem, would also not be allowed in other countries. So Baby Valentina died, and her organs could not be used to save any other babies.

Only two months after the death of Baby Valentina, in Pittsburgh, Dr. Thomas Starzl, a transplant surgeon, removed the liver from a healthy baboon and transplanted it into the heart of a man who was dying from a liver disease. The baboon, a healthy, sentient, intelligent, responsive animal, was killed immediately after the liver was taken; the patient died about two months later. No court stepped in to prevent the use of the baboon's liver. Although some animal rights groups protested against the use of a baboon in this way, most of the discussion at the time brushed over the fact that the transplant had involved the death of a baboon, and focused instead on whether this procedure could offer new hope for a large number of people needing organ transplants.

The traditional sanctity-of-life ethic forbids us to kill and take the organs of a human being who is not, and never can be, even minimally conscious; and it maintains this refusal even when the parents of the infant favor the donation of the organs. At the same time, this ethic accepts without question that we may rear baboons and chimpanzees in order to kill them and use their organs. Why does our ethic draw so sharp a distinction between human beings and all other animals? Why does species membership make such a difference to the ethics of how we may treat a being?

The Western Tradition Under Attack

MANY WRITERS HAVE DESCRIBED in detail how the Western tradition has put human beings on a pinnacle and separated them from the nonhuman animals. It was to humans that God gave dominion over the other animals; it was humans who were made in the image of God; and it was humans, and only humans, who had an immortal soul. For thousands of years, the human-centered Western tradition ruled without serious opposition. Then, in 1838, a young scientist wrote in his notebook:

> Man in his arrogance thinks himself a great work, worthy of the interposition of a deity. More humble and, I believe, true to consider him created from animals.[3]

That young scientist was, of course, Charles Darwin. It took another thirty-three years until, with the publication of *The Descent of Man,* in 1871, he was prepared to say publicly what he had written in his notebook. And when he did, he undermined the foundations of the entire Western way of thinking on the place of our species in the universe. We now knew that we, too, were animals, and had a natural origin as the other animals did. We began to see the differences between us and the nonhuman animals as differences of degree, not of kind.

For a full century, the Western view of the special status of humans did the intellectual equivalent of defying gravity: its foundations had been knocked out from under it, and yet it continued to hang there. Then a series of further waves crashed against its weakened foundations, bringing much nearer to completion the task that Darwin began. This series of waves came from several directions in quick succession.

One wave came out of the new concern with the damage we are doing to the ecosystems of our planet, to other species of animals, and to our own air and water. This gave rise to a reassessment of our attitude to the natural world. Many voices were raised in favor of dethroning human beings from their dominion and reintegrating them with the natural world.

A second wave came in the 1970s with the emergence of the animal liberation or animal rights movement, which demanded an end to "speciesism." That term, coined by the Oxford psychologist Richard Ryder in 1970 and popularized in my own book *Animal Liberation,* was used to draw a parallel between our attitudes to nonhuman animals and the attitudes of racists to those they regard as belonging to an inferior race.[4] In both cases there is an inner group that justifies its exploitation of an outer group by reference to a distinction that lacks real moral significance. While acknowledging the acceptance of a basic principle of equality among all human beings as a progressive step, animal liberationists pointed out that the idea of human equality still left most sentient creatures outside the charmed circle. If we are now able to see that the fact that a human being belongs to a different race is not a good reason for giving less consideration to the interests of that being, then why, animal liberationists asked, should the fact that a being belongs to a different species be a good reason for doing so? The animal liberation movement demanded that we go beyond a speciesist morality and give equal consideration to the interests of all beings who can feel pleasure or pain, irrespective of species.

Within a decade of its founding, the animal liberation movement had grown remarkably, influencing public debate in every area of the treatment of animals, from product testing on animals to factory farming, the use of animals for fur and in circuses, and the whaling industry. Part of the strength of the movement was its solid philosophical base. The animal liberation movement is unique among recent political movements in the extent to which its ideas and support have come from academic philosophers. This has meant that the case against speciesism has been put more rigorously than might otherwise have been the case. Many philosophers now accept that a difference of species alone cannot provide an ethically defensible basis for giving the interests of one being more consideration than another. Their challenge to our automatic assumption of species superiority has had an impact on both our thought and our practice that goes far beyond their numbers.

A third wave came from our growing knowledge of nonhuman animals, and especially of the great apes. Jane Goodall was the first human being to be so well accepted by a group of free-living chimpanzees that she could spend hundreds of hours observing them at close range in their natural habitat. Dian Fossey carried out similar studies of gorillas in Rwanda. The work of these two women led to books read by millions and started a public fascination with our nearest relatives which has continued through films and television programs. Their observations have repeatedly broken down the barriers we have erected between ourselves and other animals. People used to say that one distinguishing mark of our species was that only humans use tools. Then Goodall saw chimpanzees use sticks in order to fish for termites inside a termite nest. Other scientists reported that some seals will use rocks in order to break open shellfish, and various birds use thorns or small sticks to probe insects out of bark. So we retreated a little, and said that only humans *make* tools. This point of demarcation collapsed when Goodall reported that chimpanzees do shape their sticks, by stripping off leaves and small branches until they get the right kind of implement for the task. Since then chimpanzees have been seen to make and use a variety of other tools; one has even used one stone to turn another stone into a sharp-edged cutting tool.

Many still regarded language as the decisive difference between us and them. Early attempts to teach chimpanzees to speak failed, because they simply do not have the vocal cords needed to produce words. When two American scientists brought an infant chimpanzee called Washoe into their home and raised her exactly as if she were a deaf human child, using American Sign Language to communicate with her and with each other in her presence, they found that she learned to understand and use a large number of signs. When she matured and had an adoptive son, she taught him signs, which he then used to communicate with other chimpanzees as well as with humans. Washoe has also been filmed signing to herself when no one else is around. Gorillas and orangutans have also learned American Sign Language. Koko, a gorilla, has a vocabulary of over one thousand words and can understand a much larger number of words in spoken English as well. Chantek, an orangutan, used signs to tell lies. For example, he stole a pencil eraser, put it in his mouth, and signed "food eat," as if to say that he had swallowed it. He had really kept the eraser in his cheek, and later it was found in his bedroom in a place where he commonly hid things. Although there have been attempts to suggest that what the apes are doing is not "really" language but just a response to cues provided by human beings, the immense amount of data now accumulated, including the data on apes signing to each other or to themselves when no humans are around, now make this explanation untenable.[5]

What may prove the final blow to the traditional Western view of the distinctness of human beings is now coming from new knowledge in genetics and its implications for the way scientists classify humans and our nearest ancestors. For many years, most biologists assumed that humans evolved as a separate branch from the other great apes, including the chimpanzees and gorillas. This was a natural enough assumption, given that in many ways they look more like each other than they look like us. More recent techniques in molecular biology have enabled us to measure quite precisely the degree of genetic difference between different animals. We now know that we share 98.4 percent of our DNA with chimpanzees. This is a very slight genetic difference. It is, for example, less than that between two different species of gibbon, which are separated by 2.2 percent; or between the red-eyed and white-eyed vireos, two closely related North American bird species, the genes of which differ by 2.9 percent. More significant still is the fact that the difference between us and chimpanzees is less than the 2.3 percent that separates the DNA of chimpanzees from that of gorillas. In other words, we—not the gorillas—are the chimpanzees' nearest relatives. And all of the African apes—chimpanzees, gorillas, and humans—are more closely related to each other than any of them are to orangutans.

On the basis of this discovery, some leading scientists, among them Richard Dawkins, lecturer in zoology at the University of Oxford, and Jared Diamond, professor of physiology at the University of California, Los Angeles, have proposed that we should change the way we classify ourselves and the other African apes. As we presently classify ourselves, not only are humans a separate species, *Homo sapiens*, but we also have our own genus, *Homo,* and even our own family (the next grouping up), Hominidae. Our nearest relative, the chimpanzee, is not *Homo* but *Pan* (there are two species, *Pan troglodytes*, the common chimpanzee, and *Pan paniscus*, the pygmy chimpanzee or Gonobo), while the gorilla is a separate family, *Gorilla gorilla*, and the apes as a whole belong to the family Pongidae. We now have decisive evidence that this categorization has no basis other than the desire to separate us from other animals. All taxonomists agree that the two species of gibbon belong in the same genus, and the same is true of the red-eyed and white-eyed vireos. We are closer to the chimpanzees than the different species of gibbons, or the different species of vireos, are to each other. We are also approximately as close to the gorillas as these different species are to each other. There is only one proper conclusion to draw: we belong in the same genus as the chimpanzees and gorillas. Since the rules of naming in zoology give priority to the name that was first proposed, this means that the two species of chimpanzee should be renamed *Homo troglodytes* and *Homo paniscus*, and the gorilla, *Homo gorilla*. As Jared Diamond has put it, we are "the third chimpanzee."[6]

The Great Ape Project

ALL OF THIS MEANS that the time has come for a new idea: to extend the moral community beyond human beings to chimpanzees, gorillas, and orangutans. The "community of equals" would then embrace all the great apes, not only members of our own species. By "community of equals," Paola Cavalieri and I—the cofounders of the Great Ape Project—mean the moral community within which the most basic ethical and legal principles apply to all members. At present, in every civilized human society, only human beings are recognized as having a right to life, a right not to be tortured, and a right not to be imprisoned without due process. Internationally accepted ethical standards recognize that all humans have these basic rights—and that only humans have them. The central tenet of the Great Ape Project is that it is ethically indefensible to deny the great apes these basic rights.

Some will laugh at the idea of extending fundamental "human" rights to chimpanzees, gorillas, and orangutans. But in 1821, the House of Commons broke into howls of laughter when Richard Martin, the passionate member for Galway, proposed a law to protect horses from ill treatment, and Alderman C. Smith suggested that asses should be given protection as well. The argument from ridicule is too facile. What one generation finds ridiculous, the next accepts; and the third shudders when it looks back on what the first did.

Many will say that it is wrong to concern ourselves with extending rights to apes when so many humans are unable to enjoy the basic rights that, under various United Nations Declarations, they already hold. That human beings need better protection from murder, arbitrary imprisonment, and torture is undeniable; but why is this a reason for not doing anything about the rights of those outside our species? When I urge that developed nations ought to be giving far more to help those in danger of starvation in third world nations, I often get the reply: "Don't we have enough people in need of help right here at home?" Yes, we do, but if we waited until everything was perfect at home, we would never do anything for those farther away, where the need is often greater and the remedy more apparent. The same applies to our efforts to stop the killing, arbitrary confinement, and infliction of pain and disease on the great apes. They cannot wait until we have dealt with all abuses of human beings.

Those who know my book *Animal Liberation* may wonder why I now seem to be prepared to restrict my campaign for equality to the great apes. After all, the central thesis of *Animal Liberation* is that the basic principle of equality that entitles us to regard all human beings as equal—the principle of equal consideration of interests—ought to be applied to all beings who have interests. Since all beings capable of experiencing pleasure and pain have interests, this includes all mammals, indeed all vertebrates, and probably many invertebrates as well. The book focuses especially on the exploitation of animals in intensive farms and laboratories. Since none of the great apes are farmed and the few thousand great apes—mainly chimpanzees—in laboratories make up an insignificant proportion of the millions of animals used in research, the great apes barely rate a mention in *Animal Liberation.* What has made me narrow my focus so dramatically?

I have not changed my views about extending the basic principle of equality to all sentient beings, but this will inevitably be a long and slow process. All over the world people are involved in raising and killing sentient animals for food. The extension of the community of equals to all sentient beings will remain politically impossible for a long time to come, no matter how strong the ethical arguments for such an extension may be. By comparison, the step advocated by the Great Ape Project is relatively modest. It would stop experimentation on chimpanzees, the confinement of gorillas in sterile zoo cages, and the pathetic displays of orangutans for entertainment purposes, but on the whole it would not upset any major industry or population of voters. It seems reasonable to hope that the idea of granting basic rights to the other great apes could, over the next few years, gain strong political support on an international level. Yet to extend the community of equals would, at the same time, be a historic breakthrough in our thinking. The Great Ape Project is not an appeal to save endangered animals before they become extinct or a plea for more humane treatment. It is a call to respect the rights of individual animals in the same way that we respect the rights of human beings. This is something that has never happened before. Its achievement would make a breach in the species barrier that would, in time, make it easier to reach out to other nonhuman beings.

The Great Ape Project is also a book, now available in English, German, Italian, Spanish, and Japanese, that centers on a "Declaration on the Great Apes," setting out the position I have just outlined. It includes thirty-four essays by distinguished scientists, philosophers, and others, indicating why they support the declaration. Many of the most powerful essays come from those who have been close to great apes, either in the wild or by communicating with them through sign language, in the way I have already described. In his contribution Geza Teleki, chairman of the Washington-based Committee for Conservation and Care of Chimpanzees, describes an evening in the Gombe hills, where he had gone to study chimpanzees under the guidance of Jane Goodall. As he sat on a grassy ridge watching the sun sink over Lake Tanganyika, two adult male chimpanzees climbed the ridge from opposite sides. As they met at the top they stood upright, face to face, and clasped hands, while softly panting. Then they sat down together and joined Teleki in watching the sunset.

Such first moments reveal only a glimpse of what our nearest relatives are like. After twenty-five years of watching chimpanzees in the Mahale Mountains of Tanzania, Toshisada Nishida still observes forms of behavior that are totally new and that, as he writes, "continue to be the source of astonishment, interest and pleasure."[7] The apes who have learned sign language also continue to display previously undreamed-of capacities. Washoe, a chimpanzee, was reared from infancy by Allen and Beatrice Gardner, who taught her to use sign language. When she was five years old, they sent her away with Roger and Deborah Fouts to an institution in Oklahoma. It was eleven years before they saw Washoe again. When, after that time, they unexpectedly entered the room where she was, Washoe signed their name signs, then signed "Come Mrs. G," led Beatrice Gardner to an adjoining room, and began to play a game with her that she had not been observed playing since she left the Gardners' home.

Where does all this lead, in ethical and political terms? In our own concluding essay to *The Great Ape Project*, Paola Cavalieri and I point to an international organization that can serve as a political model for the liberation of the great apes: the Anti-Slavery Society. In the last two hundred years, human slavery has been eliminated, or virtually so, from the face of the earth. In laboratories, zoos, circuses, and elsewhere, the great apes remain the most abject of slaves. In 1991, the United States government set official minimum standards for cages for laboratory chimpanzees. The recommended cage size for permanently confining a single adult chimpanzee was 5 × 5 × 7 feet.

Can we put an end to the slavery of the great apes? Since the launch of the Great Ape Project in 1993 there have been promising signs of change in attitudes toward the nonhuman great apes. In Britain, the government has said that it will no longer allow great apes to be the subject of harmful scientific experimentation. In the United States, a National Science Foundation report on chimpanzees in laboratories recommended that—in contrast to the methods normally applied to other animals used in research—chimpanzees surplus to research requirements should not be killed but should be "retired." In New Zealand, animal welfare legislation enacted in 1999 included a clause prohibiting the use of "non-human hominoids"—that is, great apes—in research, unless the research was intended to benefit them, either individually or as a species. In hailing the legislation as a world first, the government minister responsible for its passage through parliament specifically referred to "the advanced cognitive and emotional capacity of the great apes" as the reason for the special status being accorded to them.

These are the early signs of what could become a more fundamental change. It needs to happen soon, not only for the sake of those apes now enslaved in the developed world, but also for those still living freely in their original lands. In Africa, and in Indonesia, the great apes are everywhere endangered by the clearing and burning of the forests in which they live. In Africa chimpanzees, bonobos, and gorillas face an additional threat: foreign logging companies are building roads into the forests, providing access for hunters who shoot them for "bushmeat," often to be sold at high prices in markets and restaurants hundreds of miles away. If this trade is not stopped soon, there will be very few free-living African apes left.

Extending the community of equals to the great apes is a first step toward the broader moral community that should eventually include all sentient creatures. We imagine that there is a vast gulf between us and other species. This gulf has disastrous consequences not only for the great apes, but for all animals. The Great Ape Project is not merely the champion of the one relatively small group of animals over others, but rather a bridge that will reduce this gulf, and so lead to a different attitude toward all sentient creatures.[8]

Environmental Values

FROM *The Environmental Challenge*

A RIVER TUMBLES through steep wooded valleys and rocky gorges toward the sea. The state hydroelectricity commission sees the falling water as untapped energy. Building a dam across one of the gorges would provide three years of employment for a thousand people, and longer-term employment for twenty or thirty. The dam would store enough water to ensure that the state could economically meet its energy needs for the next decade. This would encourage the establishment of energy-intensive industry in the state, thus further contributing to employment and economic growth.

The rough terrain of the river valley makes it accessible only to the reasonably fit, but it is nevertheless a favored spot for bushwalking. The river itself attracts the more daring white-water rafters. Deep in the sheltered valleys are stands of Huon pine, thousands of years old. The valleys and gorges are home to many birds and animals, including an endangered species of marsupial mouse found in only one other place in Australia. There may be other rare plants and animals as well, but no one knows, for scientists have yet to investigate the region fully.

Should the dam be built? This is one example of a situation in which we must choose between very different sets of values. The description is clearly reminiscent of the controversy over the proposed dam on the Franklin River, in Tasmania's southwest, although I have not tried to make it exact, and it should be treated as a hypothetical case.[1] Many other examples would have posed the choice between values equally well: logging virgin forests, building a paper mill that will release pollutants into coastal waters, or opening a new mine on the edge of a national park. A different set of examples would raise related, but slightly different, issues: banning the use of CFCs to prevent the depletion of the ozone layer; restricting the use of fossil fuels in an attempt to slow the greenhouse effect; mining uranium, where the issue is not the potential damage to the area around the mine but the hazards of nuclear fuel. My aim in this chapter is to explore the values that underlie debates about these decisions, and the example I have presented can serve as a point of reference for these debates. I shall focus particularly on the values at issue in controversies about the preservation of wilderness, because here the fundamentally different values of the two parties are most apparent. On controversies like water pollution and the control of greenhouse gases, the difference in values tends to be obscured by scientific debates on what is really happening, what the costs are, and what measures will be effective. When we are talking about flooding a river valley, the choice before us is starkly clear.

In general terms, we can say that those who favor building the dam are valuing employment and a higher per capita income for the state above the preservation of wilderness, of plants and animals (both common ones and members of an endangered species) and of opportunities for outdoor recreational activities. Some fundamental questions of philosophy lie behind this difference of values. In what follows, I shall assume that the values we hold may properly be subjected to rational scrutiny and criticism: they are not simply matters of taste, about which argument is futile.[2] Before we begin to scrutinize the values of those who would have the dam built and those who would not, however, let us briefly investigate the origins of modern attitudes toward the natural world.

The Western Tradition

WESTERN ATTITUDES toward nature grew out of a blend of those of the Hebrew people, as represented in the early books of the Bible, and the philosophy of ancient Greece, particularly that of Aristotle. In contrast to some other ancient traditions—for example, those of India—both the Hebrew and the Greek traditions made human beings the center of the moral universe; indeed not merely the center, but very often the entirety of the morally significant features of this world.

The biblical story of creation makes very clear the Hebrew view of the special place of human beings in the divine plan:

> And God said, Let us make man in our image, after our likeness: and let them have dominion over the fish of the sea, and over the fowl of the air, and over the cattle, and over all the earth, and over every creeping thing that creepeth upon the earth.
>
> So God created man in his own image, in the image of God created he him; male and female created he them.
>
> And God blessed them, and God said unto them, Be fruitful, and multiply, and replenish the earth, and subdue it; and have dominion over the fish of the sea and over the fowl of the air, and over every living thing that moveth upon the earth.[3]

Today Christians debate the meaning of this grant of "dominion"; and those concerned about the environment claim that it should be regarded not as a licence to humanity to do as they will with other living things, but rather as a directive to look after them, on God's behalf, and be answerable to God for the way in which they are treated.[4] There is, however, little justification in the text itself for such an interpretation; and given the example God set when he drowned almost every animal on earth in order to punish Noah for his wickedness, it is no wonder that people should think the flooding of a single river valley is nothing worth worrying about. After the flood there is a repetition of the grant of dominion in more ominous language:

> And the fear of you and the dread of you shall be upon every beast of the earth, and upon every fowl of the air, upon all that moveth upon the earth, and upon all the fishes of the sea; into your hand are they delivered.[5]

The implication is clear: to act in a way that causes fear and dread to everything that moves on the earth is not improper; it is, in fact, in accordance with a God-given decree.

The most influential early Christian thinkers had no doubts about how human dominion was to be understood. "Doth God care for oxen?" asked Paul, in the course of a discussion of an Old Testament command to rest one's ox on the sabbath, but it was only a rhetorical question—he took it for granted that the answer must be negative, and the command was to be explained in terms of some benefit to humans.[6] Augustine shared this line of thought; referring to stories in the New Testament in which Jesus destroyed a fig tree and caused a herd of pigs to drown, Augustine explained these puzzling incidents as intended to teach us that "to refrain from the killing of animals and the destroying of plants is the height of superstition."[7]

When Christianity prevailed in the Roman Empire, it also absorbed elements of the ancient Greek attitude toward the natural world. The Greek influence was entrenched in Christian philosophy by the greatest of the medieval scholastics, Thomas Aquinas, whose lifework was the melding of Christian theology with the thought of Aristotle. Aristotle regarded nature as a hierarchy in which those with less reasoning ability exist for the sake of those with more:

> Plants exist for the sake of animals, and brute beasts for the sake of man—domestic animals for his use and food, wild ones (or at any rate most of them) for food and other accessories of life, such as clothing and various tools.
>
> Since nature makes nothing purposeless or in vain, it is undeniably true that she has made all animals for the sake of man.[8]

In his own major work, *Summa Theologica*, Aquinas followed this passage from Aristotle almost word for word, adding that the position accords with God's command, as given in Genesis. In his classification of sins, Aquinas has room only for sins against God, ourselves, or our neighbor. There is no possibility of sinning against nonhuman animals or against the natural world.[9]

This was the thinking of mainstream Christianity for at least its first eighteen centuries. There were gentler spirits, certainly, like Basil, John Chrysostom, and Francis of Assisi, but for most of Christian history they have had no significant impact on the dominant tradition.[10] It is therefore worth emphasising the major features of this dominant Western tradition, because these features can serve as a point of comparison when we discuss different views of the natural environment.

According to the dominant Western tradition, the natural world exists for the benefit of human beings. God gave human beings dominion over the natural world, and God does not care how we treat it. Human beings are the only morally important members of this world. Nature itself is of no intrinsic value, and the destruction of plants and animals cannot be sinful, unless by this destruction we harm human beings.

Harsh as this tradition is, it does not rule out concern for the preservation of nature, as long as that concern can be related to human well-being. Often, of course, it can be. We could, entirely within the limits of the dominant Western tradition, oppose the mining of uranium on the argument that nuclear fuel, whether in bombs or power stations, is so hazardous to human life that the uranium is better left in the ground. Similarly, many arguments against pollution, the use of gases harmful to the ozone layer, the burning of fossil fuels, and the destruction of forests could be couched in terms of the harm to human health and welfare from the pollutants, or the changes to the climate that may occur as a result of the use of fossil fuels and the loss of forest. Since human beings need an environment in which they can thrive, the preservation of such an environment can be a value within a human-centred moral framework.

From the standpoint of a form of civilization based on growing crops and grazing animals, wilderness may seem to be a wasteland, a useless area that needs clearing in order to render it productive and valuable. There was a time when villages surrounded by farmland seemed like oases of cultivation among the deserts of forest or rough mountain slopes. Now, however, a different metaphor is more appropriate: the remnants of true wilderness left to us are like islands amid a sea of human activity that threatens to engulf them. This gives wilderness a scarcity value that provides the basis for a strong argument for preservation, even within the terms of a human-centered ethic. That argument becomes much stronger still when we take a long-term view. To this immensely important aspect of environmental values we shall now turn.

Valuing the Future

A virgin forest is the product of all the millions of years that have passed since the beginning of our planet. If it is cut down, another forest may grow up, but the continuity has been broken. The disruption in the natural life cycles of the plants and animals means that the forest will never again be as it would have been had it not been cut. The gains made from cutting the forest—employment, profits for business, export earnings, and cheaper cardboard and paper for packaging—are short-term. Once the forest is cut or drowned, however, the link with the past is gone forever. That may be regretted by every generation that succeeds us on this planet. True wilderness now has a high value because it is already scarce. In the future, and considering the world as a whole, it is bound to become scarcer still. It is for that reason that environmentalists are right to speak of wilderness as a "world heritage." It is something that we have inherited from our ancestors, and that we must preserve for our descendants if they are to have it at all.

In contrast to many more stable, tradition-oriented human societies, our modern political and cultural ethos has great difficulty in recognizing long-term values. It is notorious that politicians rarely look beyond the next election; but even if they do, they will find their economic advisers telling them that anything to be gained in the future should be discounted to such a degree as to make it easy to disregard the long-term future altogether. Economists have been taught to apply a discount rate to all future goods. In other words, a million dollars in twenty years is not worth a million dollars today, even when we allow for inflation. Economists will discount the value of the million dollars by a certain percentage, usually corresponding to real long-term interest rates. This makes economic sense, because if I had a thousand dollars today I could invest it so that it would be worth more, in real terms, in twenty years. But the use of a discount rate means that values gained one hundred years hence rank very low in comparison with values gained today, and values gained one thousand years in the future scarcely count at all. This is not because of any uncertainty about whether there will be human beings or other sentient creatures inhabiting this planet at that time, but merely because of the cumulative effect of the rate of return on money invested now. From the standpoint of the priceless and timeless values we can gain from the wilderness, however, applying a discount rate gives us the wrong answer. There are some things that, once lost, no amount of money can regain. Thus to justify the destruction of an ancient forest on the grounds that it will earn us substantial export income is problematic, even if we could invest that income and increase its value from year to year; for no matter how much we increased its value, it could never buy back the link with the past represented by the forest.

This argument does not show that there can be no justification for cutting any virgin forests, but it does mean that any such justification must take full account of the value of the forests to the generations to come in the more remote future, as well as in the more immediate future. This value will obviously be related to the particular scenic or biological significance of the forest; but as the proportion of true wilderness on the earth dwindles, every part of it becomes significant, because the opportunities for experiencing wilderness become scarce, and the likelihood that a reasonable selection of the major forms of wilderness will be preserved is reduced. Moreover it is not only the people who visit a wilderness who gain from its presence. Popular support for wilderness preservation is far broader than could be the case if only those who visited were in favor of preserving it. It seems that people like to "know that it is there" even if they never see it except on their television sets.

Can we be sure that future generations will appreciate the wilderness? Perhaps they will be happier sitting in air-conditioned shopping malls, playing computer games more sophisticated than any we can imagine. That is possible, but there are several reasons why we should not give this possibility too much weight. First, the trend has been in the opposite direction: the appreciation of wilderness has never been higher than it is today, especially among those nations that have overcome the problems of poverty and hunger and have relatively little wilderness left. Wilderness is valued as something of immense beauty, as a reservoir of scientific knowledge still to be gained, for the unique recreational opportunities that it provides, and because many people just like to know that something natural is still there, relatively untouched by modern civilization. If, as we all hope, future generations are able to provide for the basic needs of most people, we can expect that, for centuries to come, they too will value wilderness for the same reasons that we value it.

Second, arguments for preservation based on the beauty of wilderness are sometimes treated as if they were of little weight because they are "merely aesthetic." That is a mistake. We go to great lengths to preserve the artistic treasures of earlier human civilizations. It is difficult to imagine any economic gain that we would be prepared to accept as adequate compensation for the destruction of the paintings in the Louvre, for instance. How should we compare the aesthetic value of a wild river valley or a virgin forest with that of the paintings in the Louvre? Here, perhaps, judgement does become inescapably subjective, so I shall report my own experiences. I have looked at the paintings in the Louvre, and in many of the other great galleries of Europe and the United States. I think I have a reasonable sense of appreciation of the fine arts, yet I have not had, in any museum, experiences that have filled my aesthetic senses in the way that I experience when I walk in a natural setting and pause to survey the view from a rocky peak overlooking a forested valley, or sit by a stream tumbling over moss-covered boulders set among tall green tree-ferns. I do not think that I am alone in this. For many people, wilderness is the source of the greatest feelings of aesthetic appreciation; even nonreligious people tend to describe it in terms of a spiritual experience.

It may nevertheless be true that this appreciation of nature may not be shared by people living a century or two hence. But if wilderness can be the source of such deep joy and satisfaction, that would be a great loss. Moreover, whether future generations value wilderness is up to us; it is, at least, something we can influence. By our preservation of areas of wilderness, we provide an opportunity for generations to come, and by the books and films we produce, we create a culture that can be handed on to our children and their children. If we feel that a walk in the forest, with senses attuned to the appreciation of such an experience, is a more deeply rewarding way to spend a day than playing computer games, or if we feel that to carry one's food and shelter in a backpack for a week while hiking through an unspoiled natural environment will do more to develop character than watching television for an equivalent period, then we ought to encourage future generations to have a feeling for nature; if they end up preferring computer games, we shall have failed.

Finally, if we preserve intact the amount of wilderness that exists now, future generations will at least be able to have the choice of getting up from their computer games and going to see a world that has not been created by human beings. They may like to know that some parts of the world in which they are living are much as they were before we humans developed our fearsome powers of destruction. At present, there are still both wilderness and nonwilderness areas. If we destroy the wilderness, that choice is gone forever. Just as we will spend large sums to preserve cities like Venice, even though future generations conceivably may not be interested in such architectural treasures, so we should preserve wilderness even though it is possible that future generations will care little for it. Thus we will not wrong future generations, as we have been wronged by members of past generations whose thoughtless actions have deprived us of so many possibilities, like the chance of glimpsing the thylacine when walking through Tasmanian forests. We must take care not to inflict equally irreparable losses on the generations to follow us.

Thus a human-centered ethic can be the basis of powerful arguments for what we may call "environmental values." Even from the perspective of such an ethic, economic growth based on the exploitation of irreplaceable resources can be seen as something that brings gains to the present generation, and possibly the next generation or two, at a price that will be paid by every generation to come. The price to be paid by future human beings is too high. But should we limit ourselves to a human-centered ethic? We now need to consider more fundamental challenges to this traditional Western approach to environmental issues.

Is There Value Beyond the Human Species?

Although some debates about significant environmental issues can be conducted by appealing only to the long-term interests of our own species, in any serious exploration of environmental values a central issue will be whether there is anything of intrinsic value beyond human beings. To explore this question we first need to understand the notion of "intrinsic value." Something is of intrinsic value if it is good or desirable in itself; the contrast is with instrumental value, which is value as a means to some other end or purpose. Our own happiness, for example, is of intrinsic value, at least to most of us, in that we desire it for its own sake. Money, on the other hand, is only of instrumental value to us. We want it because of the things we can buy with it, but if we were marooned on a desert island, we would not want it (whereas happiness would be just as important to us on a desert island as anywhere else).

Now consider again for a moment the issue of damming the river described at the beginning of this chapter. Should the decision be made on the basis of human interests alone? If we say that it should, we shall balance the economic benefits for Tasmanians of building the dam against the loss for bushwalkers, scientists, and others, now and in the future, who value the preservation of the river in its natural state. We have already seen that because this calculation includes an indefinite number of future generations, the loss of the wild river is a much greater cost than we might at first imagine. Even so, if we are justified in arguing that the decision whether to dam the river should be made on the basis of values that include, but are not limited to, the interests of human beings, we may have much more to set against the economic benefits for Tasmanians of building the dam. We may take into account the interests of the animals who will die if the valley is drowned; we may give weight to the fact that a species may be lost, that trees that have stood for thousands of years will die, and that an entire local ecosystem will be destroyed; and we may give the preservation of the animals, the species, the trees, and the ecosystem a weight that is independent of the interests of human beings—whether economic, recreational, or scientific—in their preservation.

Here we have a fundamental moral disagreement: a disagreement about what kinds of beings ought to be considered in our moral deliberations.

Many people think that once we reach a disagreement of this kind, argument must cease. As I have already briefly indicated, I am more optimistic about the scope of rational argument in ethics. In ethics, even at a fundamental level, there are arguments that should convince any rational person. Take, as an example, a view held by one of the founders of the Western ethical tradition: Aristotle's notorious justification of slavery. Aristotle thought that captured barbarians were "living instruments"—that is, human beings who were not of intrinsic value but existed in order to serve some higher end. That end was the welfare of their Greek captors or owners. He justified this view by arguing that barbarians were less rational than Greeks and, in the hierarchy of nature, the purpose of the less rational is to serve the more rational.[11]

No one now accepts Aristotle's defense of slavery. We reject it for a variety of reasons. We would reject his assumption that non-Greeks are less rational than Greeks, although given the cultural achievements of the different groups at the time, that was by no means an absurd assumption to make. More importantly, from the moral point of view, we reject the idea that the less rational exist in order to serve the more rational. Instead we hold that all humans are equal. We regard racism and slavery based on racism as wrong because they fail to give equal consideration to the interests of all human beings. This would be true whatever the level of rationality or civilization of the slave, and therefore Aristotle's appeal to the higher rationality of the Greeks would not have justified the enslavement of non-Greeks, even if it had been true. Members of the "barbarian" tribes can feel pain, as Greeks can; they can be joyful or miserable, as Greeks can; they can suffer from separation from their families and friends, as Greeks can. To brush aside these needs so that Greeks could satisfy much more minor needs of their own was a great wrong and a blot on Greek civilization. This is something that we would expect all reasonable people to accept, as long as they can view the question from an impartial perspective and are not improperly influenced by having a personal interest in the continued existence of slavery.

Now let us return to the question of the moral status of those who are not humans. We shall consider, first, nonhuman animals. In keeping with the dominant Western tradition, many people still hold that all the nonhuman natural world has value only or predominantly insofar as it benefits human beings. A powerful objection to the dominant Western tradition turns against this tradition an extended version of the objection just made against Aristotle's justification of slavery. Many nonhuman animals are also capable of feeling pain, as humans are; they can certainly be miserable, and perhaps in some cases their lives could also be described as joyful; and members of many mammalian species can suffer from separation from their family group. Is it not therefore a blot on human civilization that we brush aside these needs of nonhuman animals to satisfy minor needs of our own?

Rejecting the dominant Western tradition in this way makes a radical difference to the value basis on which we should consider environmental policy. Into the calculations about damming the river must now go the interests of all the nonhuman animals who live in the area that will be flooded. A few may be able to move to a neighboring area that is suitable, but wilderness is not full of suitable niches awaiting an occupant; if there is territory that can sustain a native animal, it is most likely already occupied. Thus most of them will die: either they will be drowned, or they will starve.

Neither drowning nor starvation is an easy way to die, and the suffering involved in these deaths should, as we have seen, be given no less weight than we would give to an equivalent amount of suffering experienced by human beings. That, in itself, may be enough to swing the balance against building the dam. What of the fact that the animals will die, apart from the suffering that will occur in the course of dying? Are we also to weigh the deaths of nonhuman animals as equivalent to the deaths of a similar number of human beings? If so, it would seem that almost no development of any area can be justified; even industrial wastelands provide habitat for rodents who will die if the land is built upon. But the argument presented above does not require us to regard the death of a nonhuman animal as morally equivalent to the death of a human being, since humans are capable of foresight and forward planning in ways that nonhuman animals are not. This is surely relevant to the seriousness of death, which, in the case of a human being capable of planning for the future, will thwart these plans, and which thus causes a loss that is different in kind from the loss that death causes to beings incapable even of understanding that they exist over time and have a future. It is also entirely legitimate to take into account the greater sense of loss that humans feel when people close to them die; whether nonhuman animals will feel a sense of loss at the death of another animal will depend on the social habits of the species, but in most cases it is unlikely to be as prolonged, and perhaps not as deep, as the grief that humans feel. These differences between causing death to human beings and to nonhuman animals do not mean that the death of a nonhuman animal should be treated as being of no account. On the contrary, death still inflicts a loss on the animal—the loss of all its future existence, and the experiences that that future life would have contained. When a proposed dam would flood a valley and kill thousands, perhaps millions, of sentient creatures, these deaths should be given great importance in any assessment of the costs and benefits of building the dam.

Let us summarize the conclusions reached so far. We have seen that the dominant Western tradition would restrict environmental values to human interests; but this tradition is based on an indefensible prejudice in favor of the interests of our own species. We share our planet with members of other species who are also capable of feeling pain, of suffering, and of having their lives go well or badly. We are justified in regarding their experiences as having the same kind of value as our own similar experiences. The infliction of suffering on other sentient creatures should be given as much weight as we would give to the infliction of suffering on human beings. The deaths of nonhuman animals, considered independently from the suffering that often accompanies death, should also count, although not as much as the deaths of human beings.

Is There Value Beyond Sentient Beings?

Reverence for Life

The position we have now reached extends the ethic of the dominant Western tradition but in other respects is recognizably of the same type. It draws the boundary of moral consideration around all sentient creatures but leaves other living things outside that boundary. This means that if a valley is to be flooded, we should give weight to the interests of human beings, both present and future, and to the interests of the wallabies, possums, marsupial mice, and birds living there; but the drowning of the ancient forests, the possible loss of an entire species, the destruction of several complex ecosystems, and the blockage of the wild river itself and the loss of those rocky gorges, are factors to be taken into account only insofar as they adversely affect sentient creatures. Is a more radical break with the traditional position possible? Can some or all of these aspects of the flooding of the valley be shown to have intrinsic value, so that they must be taken into account independently of their effects on human beings or nonhuman animals?

To extend an ethic in a plausible way beyond sentient beings is a difficult task. An ethic based on the interests of sentient creatures is on recognizable ground. Sentient creatures have wants and desires. They prefer some states to others. We can therefore, though with much imaginative effort and no guarantee of success, form an idea of what it might be like to be that creature under particular conditions. (The question "What is it like to be a possum drowning?" at least makes sense, even if it is impossible for us to give a more precise answer than "It must be awful.") In reaching moral decisions affecting sentient creatures, we can attempt to add up the effects of different actions on all the sentient creatures affected by the alternative actions open to us. This provides us with at least some rough guide to what might be the right thing to do. But there is *nothing* that corresponds to what it is like to be a tree dying because its roots have been flooded. Once we abandon the interests of sentient creatures as our source of value, where do we find value? What is good or bad for nonsentient creatures, and why does it matter?

It might be thought that as long as we limit ourselves to living things, the answer is not too difficult to find. After all, we know what is good or bad for the plants in our garden: water, sunlight, and compost are good; extremes of heat or cold are bad. The same applies to plants in any forest or wilderness, so why can we not regard their flourishing as good in itself, independently of its usefulness to sentient creatures?

One problem here is that without conscious interests to guide us, we have no way of assessing the relative weights to be given to the flourishing of different forms of life. Is a two-thousand-year-old Huon pine more worthy of preservation than a tussock of grass? Most people will say that it is, but such a judgment seems to have more to do with our feelings of awe for the age, size, and beauty of the tree, or with the length of time it would take to replace it, than with our perception of some intrinsic value in the flourishing of an old tree that is not possessed by a young grass tussock.

If we cease talking in terms of sentience, the boundary between living and inanimate natural objects becomes more difficult to defend. Would it really be worse to cut down an old tree than to destroy a beautiful stalactite that has taken even longer to grow? On what grounds could such a judgement be made? Probably the best-known defense of an ethic that draws the boundaries of ethics around all living things is that of Albert Schweitzer. The phrase he used, "reverence for life," is often quoted; the arguments he offered in support of such a position are less well known. Here is one of the few passages in which he defended his ethic:

> Just as in my own will-to-live there is a yearning for more life, and for that mysterious exaltation of the will which is called pleasure, and terror in face of annihilation and that injury to the will-to-live which is called pain; so the same obtains in all the will-to-live around me, equally whether it can express itself to my comprehension or whether it remains unvoiced.
>
> Ethics thus consists in this, that I experience the necessity of practising the same reverence for life toward all will-to-live, as toward my own. Therein I have already the needed fundamental principle of morality. It is *good* to maintain and cherish life; it is *evil* to destroy and to check life.[12]

A similar view has been defended recently by the contemporary American philosopher Paul Taylor. In his book *Respect for Nature,* Taylor argues that every living thing is "pursuing its own good in its own unique way." Once we see this, we can see all living things "as we see ourselves" and therefore "we are ready to place the same value on their existence as we do on our own."[13]

The problem with the defenses offered by both Schweitzer and Taylor for their ethical views is that they use language metaphorically and then argue as if what they had said was literally true. We may often talk about plants "seeking" water or light so that they can survive, and this way of thinking about plants makes it easier to accept talk of their "will to live," or of their "pursuing" of their own good. Once we stop, however, to reflect on the fact that plants are not conscious and cannot engage in any intentional behavior, it is clear that all this language is metaphorical; one might just as well say that a river is pursuing its own good and striving to reach the sea, or that the "good" of a guided missile is to blow itself up along with its target. It is misleading of Schweitzer to attempt to sway us toward an ethic of reverence for all life by referring to "yearning," "exaltation," "pleasure," and "terror." Plants experience none of these.

Moreover, in the case of plants, rivers, and guided missiles, it is possible to give a purely physical explanation of what is happening; and in the absence of consciousness, there is no good reason why we should have greater respect for the physical processes that govern the growth and decay of living things than we have for those that govern nonliving things. This being so, it is at least not obvious why we should have greater reverence for a tree than for a stalactite, or for a single-celled organism than for a mountain; and we can pass silently by Taylor's even more extraordinary claim not merely that we should be ready to respect every living thing but that we should place the same value on the life of every living thing as we place on our own.

Deep Ecology

More than forty years ago the American ecologist Aldo Leopold wrote that there was a need for a "new ethic," an "ethic dealing with man's relation to land and to the animals and plants which grow upon it." His proposed "land ethic" would enlarge "the boundaries of the community to include soils, waters, plants, and animals, or collectively, the land."[14] The rise of ecological concern in the early 1970s led to a revival of this way of thinking. The Norwegian philosopher Arne Naess wrote a brief but influential article distinguishing between the "shallow" and "deep" forms of ecological thinking. Shallow ecological thinking was limited to the traditional moral framework; those who thought in this way were anxious to avoid pollution to our water supply so that we could have safe water to drink, and they sought to preserve wilderness so that people could continue to enjoy walking through it. Deep ecologists, on the other hand, wanted to preserve the integrity of the biosphere for its own sake, irrespective of the possible benefits to humans that might flow from so doing.[15] Subsequent writers who have attempted to develop some form of deep environmental theory include the Americans Bill Devall and George Sessions, and the Australians Lawrence Johnson, Val Plumwood, and Richard Sylvan.[16]

Where the reverence-for-life ethic emphasises individual living organisms, proposals for deep ecology ethics tend to take something larger as the object of value: species, ecological systems, even the biosphere as a whole. Leopold summed up the basis of his new land ethic thus: "A thing is right when it tends to preserve the integrity, stability and beauty of the biotic community. It is wrong when it tends otherwise."[17] In a paper published in 1984, Arne Naess and George Sessions set out several principles for a deep ecological ethic, beginning with the following:

1. The well-being and flourishing of human and non-human Life on Earth have value in themselves (synonyms: intrinsic value, inherent value). These values are independent of the usefulness of the non-human world for human purposes.

2. Richness and diversity of life forms contribute to the realization of these values and are also values in themselves.

3. Humans have no right to reduce this richness and diversity except to satisfy *vital* needs.[18]

Although these principles refer only to life, in the same paper Naess and Sessions say that deep ecology uses the term "biosphere in a more comprehensive non-technical way to refer also to what biologists classify as 'nonliving'; rivers (watersheds), landscapes, ecosystems." Sylvan and Plumwood also extend their ethic beyond living things, including in it an obligation "not to jeopardize the well-being of natural objects or systems without good reason."[19]

Behind this application of ethics not only to individuals but also to species and ecosystems lies some form of holism—some sense that the species or ecosystem is not just a collection of individuals but really an entity in its own right. This holism is made explicit in Lawrence Johnson's A *Morally Deep World*, probably the most detailed and carefully argued statement of the case for an ethic of deep ecology yet to appear in print. Lawrence is prepared to talk about the interests of a species in a sense that is distinct from the sum of the interests of each member of the species, and to argue that the interests of a species or an ecosystem ought to be taken into account, with individual interests, in our moral deliberations.

There is, of course, a real philosophical question about whether a species or an ecosystem can be considered the sort of individual that can have interests; and even if it can, the deep ecology ethic will face problems similar to those we identified in considering the idea of the reverence-for-life ethic. For it is necessary not merely that trees, species, and ecosystems can properly be said to have interests, but that they have morally significant interests. We saw in discussing the ethic of reverence for life that one way of establishing that an interest is morally significant is to ask what it would be like for the entity affected to have that interest unsatisfied. This works for sentient beings, but it does not work for trees, species, or ecosystems. There is nothing that corresponds to what it is like to be an ecosystem flooded by a dam. In this respect trees, ecosystems, and species are more like rocks than they are like sentient beings; so the divide between sentient and nonsentient creatures is to that extent a firmer basis for a morally important boundary than the divide between living and nonliving things or holistic entities.

If we were to adopt an ethic that attributed value to nonsentient living things, or to ecosystems as a whole, we would need to have a criterion of what made something more valuable than something else. Naess and Sessions, in common with many other deep ecologists, suggest "richness" and "diversity"; sometimes the term used is "complexity."[20] But what is it for something to be rich, diverse, or complex? Did the introduction of European birds into Australia make our birdlife richer and more diverse? If it could be shown that it did, would that make it a good thing? What if we should discover that allowing effluent from intensive farms to seep into our rivers greatly increases the number of microorganisms that live in them—thus giving rivers a different but more diverse and more complex ecosystem than they had before they were polluted. Does that make the pollution desirable?

To seek intrinsic value in diversity or complexity is a mistake. The reason why we may feel more strongly about destruction of diverse and complex ecosystems than about simpler ones (such as a field of wheat) may be the same as the reason why we feel more strongly about preservation of the ceiling of the Sistine Chapel than we do about the preservation of the ceiling of the Lecture Theatre H3 at Monash University (which is painted a uniform white). To break up Michelangelo's fresco into handy-sized chunks for sale to tourists would be lucrative, and no doubt the Vatican could put the money to good use in fighting poverty (better use than that usually made of the returns from damming rivers or clearing forests); but would it be right to do so? The objection, in both cases, is to vandalism: the destruction, for short-term gain, of something that has enduring value to sentient beings, is easy to destroy, but once destroyed can never exist again.

If the philosophical basis for a deep ecology ethic is difficult to sustain, this does not mean that the case for the preservation of wilderness is not strong. All it means is that one kind of argument—the argument for the intrinsic value of the plants, species, or ecosystems—is, at best, problematic. We are on surer ground if we confine ourselves to arguments based on the interests of sentient creatures, present and future, human and nonhuman. In my view the arguments grounded on the interests of present and future human beings, and on the interests of the sentient nonhumans who inhabit the wilderness, are quite sufficient to show that, at least in a society where no one needs to destroy wilderness in order to survive, the value of preserving the remaining significant areas of wilderness greatly exceeds the values gained by their destruction.

Saving and Taking Human Life

Famine, Affluence, and Morality

FROM *Philosophy and Public Affairs*

AS I WRITE THIS, in November 1971, people are dying in East Bengal from lack of food, shelter, and medical care. The suffering and death that are occurring there now are not inevitable, not unavoidable in any fatalistic sense of the term. Constant poverty, a cyclone, and a civil war have turned at least nine million people into destitute refugees; nevertheless, it is not beyond the capacity of the richer nations to give enough assistance to reduce any further suffering to very small proportions. The decisions and actions of human beings can prevent this kind of suffering. Unfortunately, human beings have not made the necessary decisions. At the individual level, people have, with very few exceptions, not responded to the situation in any significant way. Generally speaking, people have not given large sums to relief funds; they have not written to their parliamentary representatives demanding increased government assistance; they have not demonstrated in the streets, held symbolic fasts, or done anything else directed toward providing the refugees with the means to satisfy their essential needs. At the government level, no government has given the sort of massive aid that would enable the refugees to survive for more than a few days. Britain, for instance, has given rather more than most countries. It has, to date, given £14,750,000. For comparative purposes, Britain's share of the nonrecoverable development costs of the Anglo-French Concorde project is already in excess of

£275,000,000, and on present estimates will reach £440,000,000. The implication is that the British government values a supersonic transport more than thirty times as highly as it values the lives of the nine million refugees. Australia is another country which, on a per capita basis, is well up in the "aid to Bengal" table. Australia's aid, however, amounts to less than one-twelfth of the cost of Sydney's new opera house. The total amount given, from all sources, now stands at about £65,000,000. The estimated cost of keeping the refugees alive for one year is £464,000,000. Most of the refugees have now been in the camps for more than six months. The World Bank has said that India needs a minimum of £300,000,000 in assistance from other countries before he end of the year. It seems obvious that assistance on this scale will not be forthcoming. India will be forced to choose between letting the refugees starve or diverting funds from its own development program, which will mean that more of its own people will starve in the future.[1]

These are the essential facts about the present situation in Bengal. So far as it concerns us here, there is nothing unique about this situation except its magnitude. The Bengal emergency is just the latest and most acute of a series of major emergencies in various parts of the world, arising both from natural and from man-made causes. There are also many parts of the world in which people die from malnutrition and lack of food independent of any special emergency. I take Bengal as my example only because it is the present concern, and because the size of the problem has ensured that it has been given adequate publicity. Neither individuals nor governments can claim to be unaware of what is happening there.

What are the moral implications of a situation like this? In what follows, I shall argue that the way people in relatively affluent countries react to a situation like that in Bengal cannot be justified; indeed, the whole way we look at moral issues—our moral conceptual scheme—needs to be altered, and with it, the way of life that has come to be taken for granted in our society.

In arguing for this conclusion I will not, of course, claim to be morally neutral. I shall, however, try to argue for the moral position that I take, so that anyone who accepts certain assumptions, to be made explicit, will, I hope, accept my conclusion.

I begin with the assumption that suffering and death from lack of food, shelter, and medical care are bad. I think most people will agree about this, although one may reach the same view by different routes. I shall not argue for this view. People can hold all sorts of eccentric positions, and perhaps from some of them it would not follow that death by starvation is in itself bad. It is difficult, perhaps impossible, to refute such positions, and so for brevity I will henceforth take this assumption as accepted. Those who disagree need read no further.

My next point is this: if it is in our power to prevent something bad from happening, without thereby sacrificing anything of comparable moral importance, we ought, morally, to do it. By "without sacrificing anything of comparable moral importance" I mean without causing anything else comparably bad to happen, or doing something that is wrong in itself, or failing to promote some moral good, comparable in significance to the bad thing that we can prevent. This principle seems almost as uncontroversial as the last one. It requires us only to prevent what is bad, and not to promote what is good, and it requires this of us only when we can do it without sacrificing anything that is, from the moral point of view, comparably important. I could even, as far as the application of my argument to the Bengal emergency is concerned, qualify the point so as to make it: if it is in our power to prevent something very bad from happening, without thereby sacrificing anything morally significant, we ought, morally, to do it. An application of this principle would be as follows: if I am walking past a shallow pond and see a child drowning in it, I ought to wade in and pull the child out. This will mean getting my clothes muddy, but this is insignificant, while the death of the child would presumably be a very bad thing.

The uncontroversial appearance of the principle just stated is deceptive. If it were acted upon, even in its qualified form, our lives, our society, and our world would be fundamentally changed. For the principle takes, first, no account of proximity or distance. It makes no moral difference whether the person I can help is a neighbor's child ten yards from me or a Bengali whose name I shall never know, ten thousand miles away. Second, the principle makes no distinction between cases in which I am the only person who could possibly do anything and cases in which I am just one among millions in the same position.

I do not think I need to say much in defense of the refusal to take proximity and distance into account. The fact that a person is physically near to us, so that we have personal contact with him, may make it more likely that we *shall* assist him, but this does not show that we *ought* to help him rather than another who happens to be further away. If we accept any principle of impartiality, universalizability, equality, or whatever, we cannot discriminate against someone merely because he is far away from us (or we are far away from him). Admittedly, it is possible that we are in a better position to judge what needs to be done to help a person near to us than one far away, and perhaps also to provide the assistance we judge to be necessary. If this were the case, it would be a reason for helping those near to us first. This may once have been a justification for being more concerned with the poor in one's own town than with famine victims in India. Unfortunately for those who like to keep their moral responsibilities limited, instant communication and swift transportation have changed the situation. From the moral point of view, the development of the world into a "global village" has made an important, though still unrecognized, difference to our moral situation. Expert observers and supervisors, sent out by famine relief organizations or permanently stationed in famine-prone areas, can direct our aid to a refugee in Bengal almost as effectively as we could get it to someone in our own block. There would seem, therefore, to be no possible justification for discriminating on geographical grounds.

There may be a greater need to defend the second implication of my principle—that the fact that there are millions of other people in the same position, in respect to the Bengali refugees, as I am, does not make the situation significantly different from a situation in which I am the only person who can prevent something very bad from occurring. Again, of course, I admit that there is a psychological difference between the cases; one feels less guilty about doing nothing if one can point to others, similarly placed, who have also done nothing. Yet this can make no real difference to our moral obligations.[2] Should I consider that I am less obliged to pull the drowning child out of the pond if on looking around I see other people, no further away than I am, who have also noticed the child but are doing nothing? One has only to ask this question to see the absurdity of the view that numbers lessen obligation. It is a view that is an ideal excuse for inactivity; unfortunately most of the major evils—poverty, overpopulation, pollution—are problems in which everyone is almost equally involved.

The view that numbers do make a difference can be made plausible if stated in this way: if everyone in circumstances like mine gave £5 to the Bengal Relief Fund, there would be enough to provide food, shelter, and medical care for the refugees; there is no reason why I should give more than anyone else in the same circumstances as I am; therefore I have no obligation to give more than £5. Each premise in this argument is true, and the argument looks sound. It may convince us, unless we notice that it is based on a hypothetical premise, although the conclusion is not stated hypothetically. The argument would be sound if the conclusion were: if everyone in circumstances like mine were to give £5, I would have no obligation to give more than £5. If the conclusion were so stated, however, it would be obvious that the argument has no bearing on a situation in which it is not the case that everyone else gives £5. This, of course, is the actual situation. It is more or less certain that not everyone in circumstances like mine will give £5. So there will not be enough to provide the needed food, shelter, and medical care. Therefore by giving more than £5 I will prevent more suffering than I would if I gave just £5.

It might be thought that this argument has an absurd consequence. Since the situation appears to be that very few people are likely to give substantial amounts, it follows that I and everyone else in similar circumstances ought to give as much as possible, that is, at least up to the point at which by giving more one would begin to cause serious suffering for oneself and one's dependents—perhaps even beyond this point to the point of marginal utility, at which by giving more one would cause oneself and one's dependents as much suffering as one would prevent in Bengal. If everyone does this, however, there will be more than can be used for the benefit of the refugees, and some of the sacrifice will have been unnecessary. Thus, if everyone does what he ought to do, the result will not be as good as it would be if everyone did a little less than he ought to do, or if only some do all that they ought to do.

The paradox here arises only if we assume that the actions in question—sending money to the relief funds—are performed more or less simultaneously, and are also unexpected. For if it is to be expected that everyone is going to contribute something, then clearly each is not obliged to give as much as he would have been obliged to had others not been giving too. And if everyone is not acting more or less simultaneously, then those giving later will know how much more is needed and will have no obligation to give more than is necessary to reach this amount. To say this is not to deny the principle that people in the same circumstances have the same obligations, but to point out that the fact that others have given, or may be expected to give, is a relevant circumstance: those giving after it has become known that many others are giving and those giving before are not in the same circumstances. So the seemingly absurd consequence of the principle I have put forward can occur only if people are in error about the actual circumstances—that is, if they think they are giving when others are not, but in fact they are giving when others are. The result of everyone's doing what he really ought to do cannot be worse than the result of everyone's doing less than he ought to do, although the result of everyone's doing what he reasonably believes he ought to do could be.

If my argument so far has been sound, neither our distance from a preventable evil nor the number of other people who, in respect to that evil, are in the same situation as we are lessens our obligation to mitigate or prevent that evil. I shall therefore take as established the principle I asserted earlier. As I have already said, I need to assert it only in its qualified form: if it is in our power to prevent something very bad from happening, without thereby sacrificing anything else morally significant, we ought, morally, to do it.

The outcome of this argument is that our traditional moral categories are upset. The traditional distinction between duty and charity cannot be drawn, or at least cannot be drawn in the place we normally draw it. Giving money to the Bengal Relief Fund is regarded as an act of charity in our society. The bodies which collect money are known as "charities." These organizations see themselves in this way—if you send them a check, you will be thanked for your "generosity." Because giving money is regarded as an act of charity, it is not thought that there is anything wrong with not giving. The charitable man may be praised, but the man who is not charitable is not condemned. People do not feel in any way ashamed or guilty about spending money on new clothes or a new car instead of giving it to famine relief. (Indeed, the alternative does not occur to them.) This way of looking at the matter cannot be justified. When we buy new clothes not to keep ourselves warm but to look "well-dressed," we are not providing for any important need. We would not be sacrificing anything significant if we were to continue to wear our old clothes and give the money to famine relief. By doing so, we would be preventing another person from starving. It follows from what I have said earlier that we ought to give money away, rather than spend it on clothes which we do not need to keep us warm. To do so is not charitable or generous. Nor is it the kind of act which philosophers and theologians have called "supererogatory"—an act which it would be good to do but not wrong not to do. On the contrary, we ought to give the money away, and it is wrong not to do so.

I am not maintaining that there are no acts which are charitable, or that there are no acts which it would be good to do but not wrong not to do. It may be possible to redraw the distinction between duty and charity in some other place. All I am arguing here is that the present way of drawing the distinction, which makes it an act of charity for a man living at the level of affluence which most people in the "developed nations" enjoy to give money to save someone else from starvation, cannot be supported. It is beyond the scope of my argument to consider whether the distinction should be redrawn or abolished altogether. There would be many other possible ways of drawing the distinction—for instance, one might decide that it is good to make other people as happy as possible but not wrong not to do so.

Despite the limited nature of the revision in our moral conceptual scheme which I am proposing, the revision would, given the extent of both affluence and famine in the world today, have radical implications. These implications may lead to further objections, distinct from those I have already considered. I shall discuss two of these.

One objection to the position I have taken might be simply that it is too drastic a revision of our moral scheme. People do not ordinarily judge in the way I have suggested they should. Most people reserve their moral condemnation for those who violate some moral norm, such as the norm against taking another person's property. They do not condemn those who indulge in luxury instead of giving to famine relief. But given that I did not set out to present a morally neutral description of the way people make moral judgments, the way people do in fact judge has nothing to do with the validity of my conclusion. My conclusion follows from the principle which I advanced earlier, and unless that principle is rejected, or the arguments are shown to be unsound, I think the conclusion must stand, however strange it appears.

It might, nevertheless, be interesting to consider why our society and most other societies do judge differently from the way I have suggested they should. In a well-known article, J. O. Urmson suggests that the imperatives of duty, which tell us what we must do, as distinct from what it would be good to do but not wrong not to do, function so as to prohibit behavior that is intolerable if men are to live together in society.[3] This may explain the origin and continued existence of the present division between acts of duty and acts of charity. Moral attitudes are shaped by the needs of society, and no doubt society needs people who will observe the rules that make social existence tolerable. From the point of view of a particular society, it is essential to prevent violations of norms against killing, stealing, and so on. It is quite inessential, however, to help people outside one's own society.

If this is an explanation of our common distinction between duty and supererogation, however, it is not a justification of it. The moral point of view requires us to look beyond the interests of our own society. Previously, as I have already mentioned, this may hardly have been feasible, but it is quite feasible now. From the moral point of view, preventing the starvation of millions of people outside our society must be considered at least as pressing as upholding property norms within our society.

It has been argued by some writers, among them Sidgwick and Urmson, that we need to have a basic moral code which is not too far beyond the capacities of the ordinary man, for otherwise there will be a general breakdown of compliance with the moral code. Crudely stated, this argument suggests that if we tell people that they ought to refrain from murder and give everything they do not really need to famine relief, they will do neither, whereas if we tell them that they ought to refrain from murder and that it is good to give to famine relief but not wrong not to do so, they will at least refrain from murder. The issue here is: Where should we draw the line between conduct that is required and conduct that is good although not required, so as to get the best possible result? This would seem to be an empirical question, although a very difficult one. One objection to the Sidgwick-Urmson line of argument is that it takes insufficient account of the effect that moral standards can have on the decisions we make. Given a society in which a wealthy man who gives five percent of his income to famine relief is regarded as most generous, it is not surprising that a proposal that we all ought to give away half our incomes will be thought to be absurdly unrealistic. In a society which held that no man should have more than enough while others have less than they need, such a proposal might seem narrow-minded. What it is possible for a man to do and what he is likely to do are both, I think, very greatly influenced by what people around him are doing and expecting him to do. In any case, the possibility that by spreading the idea that we ought to be doing very much more than we are to relieve famine we shall bring about a general breakdown of moral behavior seems remote. If the stakes are an end to widespread starvation, it is worth the risk. Finally, it should be emphasized that these considerations are relevant only to the issue of what we should require from others, and not to what we ourselves ought to do.

The second objection to my attack on the present distinction between duty and charity is one which has from time to time been made against utilitarianism. It follows from some forms of utilitarian theory that we all ought, morally, to be working full time to increase the balance of happiness over misery. The position I have taken here would not lead to this conclusion in all circumstances, for if there were no bad occurrences that we could prevent without sacrificing something of comparable moral importance, my argument would have no application. Given the present conditions in many parts of the world, however, it does follow from my argument that we ought, morally, to be working full time to relieve great suffering of the sort that occurs as a result of famine or other disasters. Of course, mitigating circumstances can be adduced—for instance, that if we wear ourselves out through overwork, we shall be less effective than we would otherwise have been. Nevertheless, when all considerations of this sort have been taken into account, the conclusion remains: we ought to be preventing as much suffering as we can without sacrificing something else of comparable moral importance. This conclusion is one which we may be reluctant to face. I cannot see, though, why it should be regarded as a criticism of the position for which I have argued, rather than a criticism of our ordinary standards of behavior. Since most people are self-interested to some degree, very few of us are likely to do everything that we ought to do. It would, however, hardly be honest to take this as evidence that it is not the case that we ought to do it.

It may still be thought that my conclusions are so wildly out of line with what everyone else thinks and has always thought that there must be something wrong with the argument somewhere. In order to show that my conclusions, while certainly contrary to contemporary Western moral standards, would not have seemed so extraordinary at other times and in other places, I would like to quote a passage from a writer not normally thought of as a way-out radical, Thomas Aquinas.

> Now, according to the natural order instituted by divine providence, material goods are provided for the satisfaction of human needs. Therefore the division and appropriation of property, which proceeds from human law, must not hinder the satisfaction of man's necessity from such goods. Equally, whatever a man has in super-abundance is owed, of natural right, to the poor for their sustenance. So Ambrosius says, and it is also to be found in the *Decretum Gratiani:* "The bread which you withhold belongs to the hungry; the clothing you shut away, to the naked; and the money you bury in the earth is the redemption and freedom of the penniless."[4]

I now want to consider a number of points, more practical than philosophical, which are relevant to the application of the moral conclusion we have reached. These points challenge not the idea that we ought to be doing all we can to prevent starvation but the idea that giving away a great deal of money is the best means to this end.

It is sometimes said that overseas aid should be a government responsibility, and that therefore one ought not to give to privately run charities. Giving privately, it is said, allows the government and the noncontributing members of society to escape their responsibilities.

This argument seems to assume that the more people there are who give to privately organized famine relief funds, the less likely it is that the government will take over full responsibility for such aid. This assumption is unsupported and does not strike me as at all plausible. The opposite view—that if no one gives voluntarily, a government will assume that its citizens are uninterested in famine relief and would not wish to be forced into giving aid—seems more plausible. In any case, unless there were a definite probability that by refusing to give one would be helping to bring about massive government assistance, people who do refuse to make voluntary contributions are refusing to prevent a certain amount of suffering without being able to point to any tangible beneficial consequence of their refusal. So the onus of showing how their refusal will bring about government action is on those who refuse to give.

I do not, of course, want to dispute the contention that governments of affluent nations should be giving many times the amount of genuine, no-strings-attached aid that they are giving now. I agree, too, that giving privately is not enough, and that we ought to be campaigning actively for entirely new standards for both public and private contributions to famine relief. Indeed, I would sympathize with someone who thought that campaigning was more important than giving oneself, although I doubt whether preaching what one does not practice would be very effective. Unfortunately, for many people the idea that "it's the government's responsibility" is a reason for not giving which does not appear to entail any political action either.

Another, more serious reason for not giving to famine relief funds is that until there is effective population control, relieving famine merely postpones starvation. If we save the Bengal refugees now, others, perhaps the children of these refugees, will face starvation in a few years' time. In support of this, one may cite the now well-known facts about the population explosion and the relatively limited scope for expanded production.

This point, like the previous one, is an argument against relieving suffering that is happening now, because of a belief about what might happen in the future; it is unlike the previous point in that very good evidence can be adduced in support of this belief about the future. I will not go into the evidence here. I accept that the earth cannot support indefinitely a population rising at the present rate. This certainly poses a problem for anyone who thinks it important to prevent famine. Again, however, one could accept the argument without drawing the conclusion that it absolves one from any obligation to do anything to prevent famine. The conclusion that should be drawn is that the best means of preventing famine, in the long run, is population control. It would then follow from the position reached earlier that one ought to be doing all one can to promote population control (unless one held that all forms of population control were wrong in themselves or would have significantly bad consequences). Since there are organizations working specifically for population control, one would then support them rather than more orthodox methods of preventing famine.

A third point raised by the conclusion reached earlier relates to the question of just how much we all ought to be giving away. One possibility, which has already been mentioned, is that we ought to give until we reach the level of marginal utility—that is, the level at which, by giving more, I would cause as much suffering to myself or my dependents as I would relieve by my gift. This would mean, of course, that one would reduce oneself to very near the material circumstances of a Bengali refugee. It will be recalled that earlier I put forward both a strong and a moderate version of the principle of preventing bad occurrences. The strong version, which required us to prevent bad things from happening unless in doing so we would be sacrificing something of comparable moral significance, does seem to require reducing ourselves to the level of marginal utility. I should also say that the strong version seems to me to be the correct one. I proposed the more moderate version—that we should prevent bad occurrences unless, to do so, we had to sacrifice something morally significant—only in order to show that even on this surely undeniable principle a great change in our way of life is required. On the more moderate principle, it may not follow that we ought to reduce ourselves to the level of marginal utility, for one might hold that to reduce oneself and one's family to this level is to cause something significantly bad to happen. Whether this is so I shall not discuss, since, as I have said, I can see no good reason for holding the moderate version of the principle rather than the strong version. Even if we accepted the principle only in its moderate form, however, it should be clear that we would have to give away enough to ensure that the consumer society, dependent as it is on people spending on trivia rather than giving to famine relief, would slow down and perhaps disappear entirely. There are several reasons why this would be desirable in itself. The value and necessity of economic growth are now being questioned not only by conservationists but by economists as well.[5] There is no doubt, too, that the consumer society has had a distorting effect on the goals and purposes of its members. Yet looking at the matter purely from the point of view of overseas aid, there must be a limit to the extent to which we should deliberately slow down our economy, for it might be the case that if we gave away, say, forty percent of our gross national product, we would slow down the economy so much that in absolute terms we would be giving less than if we gave twenty-five percent of the much larger GNP that we would have if we limited our contribution to this smaller percentage.

I mention this only as an indication of the sort of factor that one would have to take into account in working out an ideal. Since Western societies generally consider one percent of the GNP an acceptable level for overseas aid, the matter is entirely academic. Nor does it affect the question of how much an individual should give in a society in which very few are giving substantial amounts.

It is sometimes said, though less often now than it used to be, that philosophers have no special role to play in public affairs, since most public issues depend primarily on an assessment of facts. On questions of fact, it is said, philosophers as such have no special expertise, and so it has been possible to engage in philosophy without committing oneself to any position on major public issues. No doubt there are some issues of social policy and foreign policy about which it can truly be said that a really expert assessment of the facts is required before taking sides or acting, but the issue of famine is surely not one of these. The facts about the existence of suffering are beyond dispute. Nor, I think, is it disputed that we can do something about it, either through orthodox methods of famine relief or through population control or both. This is therefore an issue on which philosophers are competent to take a position. The issue is one which faces everyone who has more money than he needs to support himself and his dependents, or who is in a position to take some sort of political action. These categories must include practically every teacher and student of philosophy in the universities of the Western world. If philosophy is to deal with matters that are relevant to both teachers and students, this is an issue that philosophers should discuss.

Discussion, though, is not enough. What is the point of relating philosophy to public (and personal) affairs if we do not take our conclusions seriously? In this instance, taking our conclusion seriously means acting upon it. The philosopher will not find it any easier than anyone else to alter his attitudes and way of life to the extent that, if I am right, is involved in doing everything that we ought to be doing. At the very least, though, one can make a start. The philosopher who does so will have to sacrifice some of the benefits of the consumer society, but he can find compensation in the satisfaction of a way of life in which theory and practice, if not yet in harmony, are at least coming together.

The Singer Solution to World Poverty

FROM *The New York Times Magazine*

IN THE BRAZILIAN FILM "Central Station," Dora is a retired schoolteacher who makes ends meet by sitting at the station writing letters for illiterate people. Suddenly she has an opportunity to pocket $1,000. All she has to do is persuade a homeless 9-year-old boy to follow her to an address she has been given. (She is told he will be adopted by wealthy foreigners.) She delivers the boy, gets the money, spends some of it on a television set, and settles down to enjoy her new acquisition. Her neighbor spoils the fun, however, by telling her that the boy was too old to be adopted—he will be killed and his organs sold for transplantation. Perhaps Dora knew this all along, but after her neighbor's plain speaking, she spends a troubled night. In the morning Dora resolves to take the boy back.

Suppose Dora had told her neighbor that it is a tough world, other people have nice new TVs too, and if selling the kid is the only way she can get one, well, he was only a street kid. She would then have become, in the eyes of the audience, a monster. She redeems herself only by being prepared to bear considerable risks to save the boy.

At the end of the movie, in cinemas in the affluent nations of the world, people who would have been quick to condemn Dora if she had not rescued the boy go home to places far more comfortable than her apartment. In fact, the average family in the United States spends almost one-third of its income on things that are no more necessary to them than Dora's new TV was to her. Going out to nice restaurants, buying new clothes because the old ones are no longer stylish, vacationing at beach resorts—so much of our income is spent on things not essential to the preservation of our lives and health. Donated to one of a number of charitable agencies, that money could mean the difference between life and death for children in need.

All of which raises a question: In the end, what is the ethical distinction between a Brazilian who sells a homeless child to organ peddlers and an American who already has a TV and upgrades to a better one—knowing that the money could be donated to an organization that would use it to save the lives of kids in need?

Of course, there are several differences between the two situations that could support different moral judgments about them. For one thing, to be able to consign a child to death when he is standing right in front of you takes a chilling kind of heartlessness; it is much easier to ignore an appeal for money to help children you will never meet. Yet for a utilitarian philosopher like myself—that is, one who judges whether acts are right or wrong by their consequences—if the upshot of the American's failure to donate the money is that one more kid dies on the streets of a Brazilian city, then it is, in some sense, just as bad as selling the kid to the organ peddlers. But one doesn't need to embrace my utilitarian ethic to see that, at the very least, there is a troubling incongruity in being so quick to condemn Dora for taking the child to the organ peddlers while, at the same time, not regarding the American consumer's behavior as raising a serious moral issue.

IN HIS 1996 BOOK, *Living High and Letting Die*, the New York University philosopher Peter Unger presented an ingenious series of imaginary examples designed to probe our intuitions about whether it is wrong to live well without giving substantial amounts of money to help people who are hungry, malnourished, or dying from easily treatable illnesses like diarrhea. Here's my paraphrase of one of these examples:

Bob is close to retirement. He has invested most of his savings in a very rare and valuable old car, a Bugatti, which he has not been able to insure. The Bugatti is his pride and joy. In addition to the pleasure he gets from driving and caring for his car, Bob knows that its rising market value means that he will always be able to sell it and live comfortably after retirement. One day when Bob is out for a drive, he parks the Bugatti near the end of a railway siding and goes for a walk up the track. As he does so, he sees that a runaway train, with no one aboard, is running down the railway track. Looking farther down the track, he sees the small figure of a child very likely to be killed by the runaway train. He can't stop the train and the child is too far away to warn of the danger, but he can throw a switch that will divert the train down the siding where his Bugatti is parked. Then nobody will be killed—but the train will destroy his Bugatti. Thinking of his joy in owning the car and the financial security it represents, Bob decides not to throw the switch. The child is killed. For many years to come, Bob enjoys owning his Bugatti and the financial security it represents.

Bob's conduct, most of us will immediately respond, was gravely wrong. Unger agrees. But then he reminds us that we, too, have opportunities to save the lives of children. We can give to organizations like Unicef or Oxfam America. How much would we have to give one of these organizations to have a high probability of saving the life of a child threatened by easily preventable diseases? (I do not believe that children are more worth saving than adults, but since no one can argue that children have brought their poverty on themselves, focusing on them simplifies the issues.) Unger called up some experts and used the information they provided to offer some plausible estimates that include the cost of raising money, administrative expenses, and the cost of delivering aid where it is most needed. By his calculation, $200 in donations would help transform a sickly 2-year-old into a healthy 6-year-old—offering safe passage through childhood's most dangerous years. To show how practical philosophical argument can be, Unger even tells his readers that they can easily donate funds by using their credit card and calling one of these toll-free numbers: (800) 367-5437 for Unicef; (800) 693-2687 for Oxfam America.

Now you, too, have the information you need to save a child's life. How should you judge yourself if you don't do it? Think again about Bob and his Bugatti. Unlike Dora, Bob did not have to look into the eyes of the child he was sacrificing for his own material comfort. The child was a complete stranger to him and too far away to relate to in an intimate, personal way. Unlike Dora, too, he did not mislead the child or initiate the chain of events imperiling him. In all these respects, Bob's situation resembles that of people able but unwilling to donate to overseas aid and differs from Dora's situation.

If you still think that it was very wrong of Bob not to throw the switch that would have diverted the train and saved the child's life, then it is hard to see how you could deny that it is also very wrong not to send money to one of the organizations listed above. Unless, that is, there is some morally important difference between the two situations that I have overlooked.

Is it the practical uncertainties about whether aid will really reach the people who need it? Nobody who knows the world of overseas aid can doubt that such uncertainties exist. But Unger's figure of $200 to save a child's life was reached after he had made conservative assumptions about the proportion of the money donated that will actually reach its target.

One genuine difference between Bob and those who can afford to donate to overseas aid organizations but don't is that only Bob can save the child on the tracks, whereas there are hundreds of millions of people who can give $200 to overseas aid organizations. The problem is that most of them aren't doing it. Does this mean that it is all right for you not to do it?

Suppose that there were more owners of priceless vintage cars—Carol, Dave, Emma, Fred, and so on, down to Ziggy—all in exactly the same situation as Bob, with their own siding and their own switch, all sacrificing the child in order to preserve their own cherished car. Would that make it all right for Bob to do the same? To answer this question affirmatively is to endorse follow-the-crowd ethics—the kind of ethics that led many Germans to look away when the Nazi atrocities were being committed. We do not excuse them because others were behaving no better.

We seem to lack a sound basis for drawing a clear moral line between Bob's situation and that of any reader of this article with $200 to spare who does not donate it to an overseas aid agency. These readers seem to be acting at least as badly as Bob was acting when he chose to let the runaway train hurtle toward the unsuspecting child. In the light of this conclusion, I trust that many readers will reach for the phone and donate that $200. Perhaps you should do it before reading further.

NOW THAT YOU have distinguished yourself morally from people who put their vintage cars ahead of a child's life, how about treating yourself and your partner to dinner at your favorite restaurant? But wait. The money you will spend at the restaurant could also help save the lives of children overseas! True, you weren't planning to blow $200 tonight, but if you were to give up dining out just for one month, you would easily save that amount. And what is one month's dining out, compared with a child's life? There's the rub. Since there are a lot of desperately needy children in the world, there will always be another child whose life you could save for another $200. Are you therefore obliged to keep giving until you have nothing left? At what point can you stop?

Hypothetical examples can easily become farcical. Consider Bob. How far past losing the Bugatti should he go? Imagine that Bob had got his foot stuck in the track of the siding, and if he diverted the train, then before it rammed the car it would also amputate his big toe. Should he still throw the switch? What if it would amputate his foot? His entire leg?

As absurd as the Bugatti scenario gets when pushed to extremes, the point it raises is a serious one: only when the sacrifices become very significant indeed would most people be prepared to say that Bob does nothing wrong when he decides not to throw the switch. Of course, most people could be wrong; we can't decide moral issues by taking opinion polls. But consider for yourself the level of sacrifice that you would demand of Bob, and then think about how much money you would have to give away in order to make a sacrifice that is roughly equal to that. It's almost certainly much, much more than $200. For most middle-class Americans, it could easily be more like $200,000.

Isn't it counterproductive to ask people to do so much? Don't we run the risk that many will shrug their shoulders and say that morality, so conceived, is fine for saints but not for them? I accept that we are unlikely to see, in the near or even medium-term future, a world in which it is normal for wealthy Americans to give the bulk of their wealth to strangers. When it comes to praising or blaming people for what they do, we tend to use a standard that is relative to some conception of normal behavior. Comfortably off Americans who give, say, 10 percent of their income to overseas aid organizations are so far ahead of most of their equally comfortable fellow citizens that I wouldn't go out of my way to chastise them for not doing more. Nevertheless, they should be doing much more, and they are in no position to criticize Bob for failing to make the much greater sacrifice of his Bugatti.

At this point various objections may crop up. Someone may say: "If every citizen living in the affluent nations contributed his or her share, I wouldn't have to make such a drastic sacrifice, because long before such levels were reached, the resources would have been there to save the lives of all those children dying from lack of food or medical care. So why should I give more than my fair share?" Another, related, objection is that the government ought to increase its overseas aid allocations, since that would spread the burden more equitably across all taxpayers.

Yet the question of how much we ought to give is a matter to be decided in the real world—and that, sadly, is a world in which we know that most people do not, and in the immediate future will not, give substantial amounts to overseas aid agencies. We know, too, that at least in the next year, the United States government is not going to meet even the very modest target, recommended by the United Nations, of 0.7 percent of gross national product; at the moment it lags far below that, at 0.09 percent, not even half of Japan's 0.22 percent or a tenth of Denmark's 0.97 percent. Thus, we know that the money we can give beyond that theoretical "fair share" is still going to save lives that would otherwise be lost. While the idea that no one need do more than his or her fair share is a powerful one, should it prevail if we know that others are not doing their fair share and that children will die preventable deaths unless we do more than our fair share? That would be taking fairness too far.

Thus, this ground for limiting how much we ought to give also fails. In the world as it is now, I can see no escape from the conclusion that each one of us with wealth surplus to his or her essential needs should be giving most of it to help people suffering from poverty so dire as to be life-threatening. That's right: I'm saying that you shouldn't buy that new car, take that cruise, redecorate the house, or get that pricey new suit. After all, a $1,000 suit could save five children's lives.

So how does my philosophy break down in dollars and cents? An American household with an income of $50,000 spends around $30,000 annually on necessities, according to the Conference Board, a nonprofit economic research organization. Therefore, for a household bringing in $50,000 a year, donations to help the world's poor should be as close as possible to $20,000. The $30,000 required for necessities holds for higher incomes as well. So a household making $100,000 could write a yearly check for $70,000. Again, the formula is simple: whatever money you're spending on luxuries, not necessities, should be given away.

Now, evolutionary psychologists tell us that human nature just isn't sufficiently altruistic to make it plausible that many people will sacrifice so much for strangers. On the facts of human nature, they might be right, but they would be wrong to draw a moral conclusion from those facts. If it is the case that we ought to do things that, predictably, most of us won't do, then let's face that fact head-on. Then, if we value the life of a child more than going to fancy restaurants, the next time we dine out we will know that we could have done something better with our money. If that makes living a morally decent life extremely arduous, well, then that is the way things are. If we don't do it, then we should at least know that we are failing to live a morally decent life—not because it is good to wallow in guilt but because knowing where we should be going is the first step toward heading in that direction.

When Bob first grasped the dilemma that faced him as he stood by that railway switch, he must have thought how extraordinarily unlucky he was to be placed in a situation in which he must choose between the life of an innocent child and the sacrifice of most of his savings. But he was not unlucky at all. We are all in that situation.

What's Wrong with Killing?

FROM *Practical Ethics*

Human Life

PEOPLE OFTEN SAY that life is sacred. They almost never mean what they say. They do not mean, as their words seem to imply, that life itself is sacred. If they did, killing a pig or pulling up a cabbage would be as abhorrent to them as the murder of a human being. When people say that life is sacred, it is human life they have in mind. But why should human life have special value?

In discussing the doctrine of the sanctity of human life I shall not take the term "sanctity" in a specifically religious sense. The doctrine may well have a religious origin, as I shall suggest later in this chapter, but it is now part of a broadly secular ethic, and it is as part of this secular ethic that it is most influential today. Nor shall I take the doctrine as maintaining that it is always wrong to take human life, for this would imply absolute pacifism, and there are many supporters of the sanctity of human life who concede that we may kill in self-defense. We may take the doctrine of the sanctity of human life to be no more than a way of saying that human life has some special value, a value quite distinct from the value of the lives of other living things.

The view that human life has unique value is deeply rooted in our society and is enshrined in our law. To see how far it can be taken, I recommend a remarkable book: *The Long Dying of Baby Andrew*, by Robert and Peggy Stinson. In December 1976 Peggy Stinson, a Pennsylvania schoolteacher, was twenty-four weeks pregnant when she went into premature labor. The baby, whom Robert and Peggy named Andrew, was marginally viable. Despite a firm statement from both parents that they wanted "no heroics," the doctors in charge of their child used all the technology of modern medicine to keep him alive for nearly six months.[1] Andrew had periodic fits. Toward the end of that period, it was clear that if he survived at all, he would be seriously and permanently impaired. Andrew was also suffering considerably: at one point his doctor told the Stinsons that it must "hurt like hell" every time Andrew drew a breath. Andrew's treatment cost $104,000, at 1977 cost levels—today it could easily be three times that, for intensive care for extremely premature babies costs at least $1,500 per day.

Andrew Stinson was kept alive, against the wishes of his parents, at a substantial financial cost, notwithstanding evident suffering, and despite the fact that, after a certain point it was clear that he would never be able to live an independent life, or to think and talk in the way that most humans do. Whether such treatment of an infant human being is or is not the right thing to do, it makes a striking contrast with the casual way in which we take the lives of stray dogs, experimental monkeys, and beef cattle. What justifies the difference?

In every society known to us there has been some prohibition on the taking of life. Presumably no society can survive if it allows its members to kill one another without restriction. Precisely who is protected, however, is a matter on which societies have differed. In many tribal societies the only serious offence is to kill an innocent member of the tribe itself—members of other tribes may be killed with impunity. In more sophisticated nation-states protection has generally extended to all within the nation's territorial boundaries, although there have been cases—like slave-owning states—in which a minority was excluded. Nowadays most societies agree, in theory if not in practice, that, apart from special cases like self-defense, war, possibly capital punishment, and one or two other doubtful areas, it is wrong to kill human beings irrespective of their race, religion, class, or nationality. The moral inadequacy of narrower principles, limiting the respect for life to a tribe, race, or nation, is taken for granted; but the argument against speciesism must raise doubts about whether the boundary of our species marks a more defensible limit to the protected circle.[2]

At this point we should pause to ask what we mean by terms like "human life" or "human being." These terms figure prominently in debates about, for example, abortion. "Is the fetus a human being?" is often taken as the crucial question in the abortion debate; but unless we think carefully about these terms such questions cannot be answered.

It is possible to give "human being" a precise meaning. We can use it as equivalent to "member of the species *Homo sapiens*." Whether a being is a member of a given species is something that can be determined scientifically, by an examination of the nature of the chromosomes in the cells of living organisms. In this sense there is no doubt that from the first moments of its existence an embryo conceived from human sperm and eggs is a human being; and the same is true of the most profoundly and irreparably intellectually disabled human being, even of an infant who is born anencephalic—literally, without a brain.

There is another use of the term "human," one proposed by Joseph Fletcher, a Protestant theologian and a prolific writer on ethical issues.[3] Fletcher has compiled a list of what he calls "indicators of humanhood" that includes the following: self-awareness, self-control, a sense of the future, a sense of the past, the capacity to relate to others, concern for others, communication, and curiosity. This is the sense of the term that we have in mind when we praise someone by saying that she is "a real human being" or shows "truly human qualities." In saying this we are not, of course, referring to the person's membership in the species *Homo sapiens*, which as a matter of biological fact is rarely in doubt; we are implying that human beings characteristically possess certain qualities, and this person possesses them to a high degree.

These two senses of "human being" overlap but do not coincide. The embryo, the later fetus, the profoundly intellectually disabled child, even the newborn infant—all are indisputably members of the species *Homo sapiens*, but none are self-aware, have a sense of the future, or have the capacity to relate to others. Hence the choice between the two senses can make an important difference to how we answer questions like "Is the fetus a human being?"

When choosing which words to use in a situation like this, we should choose terms that will enable us to express our meaning clearly and that do not prejudge the answer to substantive questions. To stipulate that we shall use "human" in, say, the first of the two senses just described, and that therefore the fetus is a human being and abortion is immoral, would not do. Nor would it be any better to choose the second sense and argue on this basis that abortion is acceptable. The morality of abortion is a substantive issue, the answer to which cannot depend on a stipulation about how we shall use words. In order to avoid begging any questions, and to make my meaning clear, I shall for the moment put aside the tricky term "human" and substitute two different terms, corresponding to the two different senses of "human." For the first sense, the biological sense, I shall simply use the cumbersome but precise expression "member of the species *Homo sapiens*" while for the second sense I shall use the term "person."

This use of "person" is itself, unfortunately, liable to mislead, since "person" is often used as if it meant the same as "human being." Yet the terms are not equivalent; there could be a person who is not a member of our species. There could also be members of our species who are not persons. The word "person" has its origin in the Latin term for a mask worn by an actor in classical drama. By putting on masks the actors signified that they were acting a role. Subsequently "person" came to mean one who plays a role in life, one who is an agent. According to the *Oxford Dictionary*, one of the current meanings of the term is "a self-conscious or rational being." This sense has impeccable philosophical precedents. John Locke defines a person as "A thinking intelligent being that has reason and reflection and can consider itself as itself, the same thinking thing, in different times and places."[4]

This definition makes "person" close to what Fletcher meant by "human," except that it selects two crucial characteristics—rationality and self-consciousness—as the core of the concept. Quite possibly Fletcher would agree that these two are central, and the others more or less follow from them. In any case, I propose to use "person," in the sense of a rational and self-conscious being, to capture those elements of the popular sense of "human being" that are not covered by "member of the species *Homo sapiens*."

The Value of the Life of Members of the Species Homo Sapiens

With the clarification gained by our terminological interlude, and the argument against speciesism to draw upon, this section can be very brief. The wrongness of inflicting pain on a being cannot depend on the being's species, nor can the wrongness of killing it. The biological facts upon which the boundary of our species is drawn do not have moral significance. To give preference to the life of a being simply because that being is a member of our species would put us in the same position as racists who give preference to those who are members of their race.

To those who have read the preceding chapters of this book, this conclusion may seem obvious, for we have worked toward it gradually; but it differs strikingly from the prevailing attitude in our society, which as we have seen treats as sacred the lives of all members of our species. How is it that our society should have come to accept a view that bears up so poorly under critical scrutiny? A short historical digression may help to explain.

If we go back to the origins of Western civilization, to Greek or Roman times, we find that membership in *Homo sapiens* was not sufficient to guarantee that one's life would be protected. There was no respect for the lives of slaves or other "barbarians"; and even among the Greeks and Romans themselves, infants had no automatic right to life. Greeks and Romans killed deformed or weak infants by exposing them to the elements on a hilltop. Plato and Aristotle thought that the state should enforce the killing of deformed infants.[5] The celebrated legislative codes said to have been drawn up by Lycurgus and Solon contained similar provisions. In this period it was thought better to end a life that had begun inauspiciously than to attempt to prolong that life, with all the problems it might bring.

Our present attitudes date from the coming of Christianity.[6] There was a specific theological motivation for the Christian insistence on the importance of species membership: the belief that all born of human parents are immortal and destined for an eternity of bliss or for everlasting torment. With this belief, the killing of *Homo sapiens* took on a fearful significance, since it consigned a being to his or her eternal fate. A second Christian doctrine that led to the same conclusion was the belief that since we are created by God, we are his property, and to kill a human being is to usurp God's right to decide when we shall live and when we shall die. As Thomas Aquinas put it, taking a human life is a sin against God in the same way that killing a slave would be a sin against the master to whom the slave belonged.[7] Nonhuman animals, on the other hand, were believed to have been placed by God under man's dominion, as recorded in the Bible (Genesis 1:29 and 9:1–3). Hence humans could kill nonhuman animals as they pleased, as long as the animals were not the property of another.

During the centuries of Christian domination of European thought the ethical attitudes based on these doctrines became part of the unquestioned moral orthodoxy of European civilization. Today the doctrines are no longer generally accepted, but the ethical attitudes to which they gave rise fit in with the deep-seated Western belief in the uniqueness and special privileges of our species, and have survived. Now that we are reassessing our speciesist view of nature, however, it is also time to reassess our belief in the sanctity of the lives of members of our species.

The Value of a Person's Life

We have broken down the doctrine of the sanctity of human life into two separate claims, one that there is special value in the life of a member of our species, and the other that there is special value in the life of a person. We have seen that the former claim cannot be defended. What of the latter? Is there special value in the life of a rational and self-conscious being, as distinct from a being that is merely sentient?

One line of argument for answering this question affirmatively runs as follows. A self-conscious being is aware of itself as a distinct entity, with a past and a future. (This, remember, was Locke's criterion for being a person.) A being aware of itself in this way will be capable of having desires about its own future. For example, a professor of philosophy may hope to write a book demonstrating the objective nature of ethics; a student may look forward to graduating; a child may want to go for a ride in an airplane. To take the lives of any of these people, without their consent, is to thwart their desires for the future. Killing a snail or a day-old infant does not thwart any desires of this kind, because snails and newborn infants are incapable of having such desires.

It may be said that when a person is killed we are not left with a thwarted desire in the same sense in which I have a thwarted desire when I am hiking through dry country and, pausing to ease my thirst, discover a hole in my water bottle. In this case I have a desire that I cannot fulfil, and I feel frustration and discomfort because of the continuing and unsatisfied desire for water. When I am killed the desires I have for the future do not continue after my death, and I do not suffer from their nonfulfilment. But does this mean that preventing the fulfilment of these desires does not matter?

Classical utilitarianism, as expounded by the founding father of utilitarianism, Jeremy Bentham, and refined by later philosophers like John Stuart Mill and Henry Sidgwick, judges actions by their tendency to maximise pleasure or happiness and minimize pain or unhappiness. Terms like "pleasure" and "happiness" lack precision, but it is clear that they refer to something that is experienced or felt—in other words, to states of consciousness. According to classical utilitarianism, therefore, there is no direct significance in the fact that desires for the future go unfulfilled when people die. If you die instantaneously, whether you have any desires for the future makes no difference to the amount of pleasure or pain you experience. Thus for the classical utilitarian the status of "person" is not *directly* relevant to the wrongness of killing.

Indirectly, however, being a person may be important for the classical utilitarian. Its importance arises in the following manner. If I am a person, I have a conception of myself. I know that I have a future. I also know that my future existence could be cut short. If I think that this is likely to happen at any moment, my present existence will be fraught with anxiety and will presumably be less enjoyable than if I do not think it is likely to happen for some time. If I learn that people like myself are very rarely killed, I will worry less. Hence the classical utilitarian can defend a prohibition on killing persons on the *indirect* ground that it will increase the happiness of people who would otherwise worry that they might be killed. I call this an *indirect* ground because it refers not to any direct wrong done to the person killed but rather to a consequence of it for other people. There is, of course, something odd about objecting to murder, not because of the wrong done to the victim, but because of the effect that the murder will have on others. One has to be a tough-minded classical utilitarian to be untroubled by this oddness. (Remember, though, that we are now considering only what is *especially* wrong about killing a *person*. The classical utilitarian can still regard killing as wrong because it eliminates the happiness that the victim would have experienced had she lived. This objection to murder will apply to any being likely to have a happy future, irrespective of whether the being is a person.) For present purposes, however, the main point is that this indirect ground does provide a reason for taking the killing of a person, under certain conditions, more seriously than the killing of a nonpersonal being. If a being is incapable of conceiving of itself as existing over time, we need not take into account the possibility of its worrying about the prospect of its future existence being cut short. It can't worry about this, for it has no conception of its own future.

I said that the indirect classical utilitarian reason for taking the killing of a person more seriously than the killing of a nonperson holds "under certain conditions." The most obvious of these conditions is that the killing of the person may become known to other persons, who derive from this knowledge a more gloomy estimate of their own chances of living to a ripe old age or simply become fearful of being murdered. It is of course possible that a person could be killed in complete secrecy, so that no one else knew a murder had been committed. Then this indirect reason against killing would not apply.

To this last point, however, a qualification must be made. In the circumstances described in the preceding paragraph, the indirect classical utilitarian reason against killing would not apply *in so far as we judge this individual case.* There is something to be said, however, against applying utilitarianism only or primarily at the level of each individual case. It may be that in the long run, we will achieve better results—greater overall happiness—if we urge people not to judge each individual action by the standard of utility, but instead to think along the lines of some broad principles that will cover all or virtually all of the situations that they are likely to encounter.

Several reasons have been offered in support of this approach. R. M. Hare has suggested a useful distinction between two levels of moral reasoning: the intuitive and the critical.[8] To consider, in theory, the possible circumstances in which one might maximise utility by secretly killing someone who wants to go on living is to reason at the critical level. It can be interesting and helpful to our understanding of ethical theory to think about such unusual hypothetical cases, as philosophers or just as reflective, self-critical people. Everyday moral thinking, however, must be more intuitive. In real life we usually cannot foresee all the complexities of our choices. It is simply not practical to try to calculate the consequences, in advance, of every choice we make. Even if we were to limit ourselves to the more significant choices, there would be a danger that in many cases we would be calculating in less than ideal circumstances. We could be hurried or flustered. We might be feeling angry, or hurt, or competitive. Our thoughts could be colored by greed, or sexual desire, or thoughts of vengeance. Our own interests, or the interests of those we love, might be at stake. Or we might just not be very good at thinking about such complicated issues as the likely consequences of a significant choice. For all these reasons, Hare suggests, it will be better if, for our everyday ethical life, we adopt some broad ethical principles and do not deviate from them. These principles should include those that experience has shown, over the centuries, to be generally conducive to producing the best consequences: and in Hare's view that would include many of the standard moral principles, for example, telling the truth, keeping promises, not harming others, and so on. Respecting the lives of people who want to go on living would presumably be among these principles. Even though, at the critical level, we can conceive of circumstances in which better consequences would flow from acting against one or more of these principles, people will do better on the whole if they stick to the principles than if they do not.

On this view, soundly chosen intuitive moral principles should be like a good tennis coach's instructions to a player. The instructions are given with an eye to what will pay off most of the time; they are a guide to playing "percentage tennis." Occasionally an individual player might go for a freak shot and pull off a winner that has everyone applauding; but if the coach is any good at all, deviations from the instructions laid down will, more often than not, lose. So it is better to put the thought of going for those freak shots out of one's mind. Similarly, if we are guided by a set of well-chosen intuitive principles, we may do better if we do not attempt to calculate the consequences of each significant moral choice we must make, but instead consider what principles apply to it, and act accordingly. Perhaps very occasionally we will find ourselves in circumstances in which it is absolutely plain that departing from the principles will produce a much better result than we will obtain by sticking to them, and then we may be justified in making the departure. But for most of us, most of the time, such circumstances will not arise and can be excluded from our thinking. Therefore even though at the critical level the classical utilitarian must concede the possibility of cases in which it would be better not to respect a person's desire to continue living, because the person could be killed in complete secrecy and a great deal of unalleviated misery could thereby be prevented, this kind of thinking has no place at the intuitive level that should guide our everyday actions. So, at least, a classical utilitarian can argue.

That is, I think, the gist of what the classical utilitarian would say about the distinction between killing a person and killing some other type of being. There is, however, another version of utilitarianism that gives greater weight to the distinction. This other version of utilitarianism judges actions, not by their tendency to maximize pleasure or minimize pain, but by the extent to which they accord with the preferences of any beings affected by the action or its consequences. This version of utilitarianism is known as "preference utilitarianism." It is preference utilitarianism, rather than classical utilitarianism, that we reach by universalizing our own interests in the manner described in the opening chapter of this book—if, that is, we make the plausible move of taking a person's interests to be what, on balance and after reflection on all the relevant facts, a person prefers.

According to preference utilitarianism, an action contrary to the preference of any being is, unless this preference is outweighed by contrary preferences, wrong. Killing a person who prefers to continue living is therefore wrong, other things being equal. That the victims are not around after the act to lament the fact that their preferences have been disregarded is irrelevant. The wrong is done when the preference is thwarted.

For preference utilitarians, taking the life of a person will normally be worse than taking the life of some other being, since persons are highly future-oriented in their preferences. To kill a person is therefore, normally, to violate not just one preference but a wide range of the most central and significant preferences a being can have. Very often, it will make nonsense of everything that the victim has been trying to do in the past days, months, or even years. In contrast, beings who cannot see themselves as entities with a future cannot have any preferences about their own future existence. This is not to deny that such beings might struggle against a situation in which their lives are in danger, as a fish struggles to get free of the barbed hook in its mouth; but this indicates no more than a preference for the cessation of a state of affairs that is perceived as painful or frightening. A struggle against danger and pain does not suggest that fish are capable of preferring their own future existence to nonexistence. The behavior of a fish on a hook suggests a reason for not killing fish by that method, but does not in itself suggest a preference utilitarian reason against killing fish by a method that brings about death instantly, without first causing pain or distress. (Again, remember that we are here considering what is especially wrong about killing a person; I am not saying that there are never any preference utilitarian reasons against killing conscious beings who are not persons.)

Does a Person Have a Right to Life?

Although preference utilitarianism does provide a direct reason for not killing a person, some may find the reason—even when coupled with the important indirect reasons that any form of utilitarianism will take into account—not sufficiently stringent. Even for preference utilitarianism, the wrong done to the person killed is merely one factor to be taken into account, and the preference of the victim could sometimes be outweighed by the preferences of others. Some say that the prohibition on killing people is more absolute than this kind of utilitarian calculation implies. Our life, we feel, is something to which we have a *right*, and rights are not to be traded off against the preferences or pleasures of others.

I am not convinced that the notion of a moral right is a helpful or meaningful one, except when it is used as a shorthand way of referring to more fundamental moral considerations. Nevertheless, since the idea that we have a "right to life" is a popular one, it is worth asking whether there are grounds for attributing rights to life to persons, as distinct from other living beings.

Michael Tooley, a contemporary American philosopher, has argued that the only beings who have a right to life are those who can conceive of themselves as distinct entities existing over time—in other words, persons, as we have used the term. His argument is based on the claim that there is a conceptual connection between the desires a being is capable of having and the rights that the being can be said to have. As Tooley puts it:

> The basic intuition is that a right is something that can be violated and that, in general, to violate an individual's right to something is to frustrate the corresponding desire. Suppose, for example, that you own a car. Then I am under a prima facie obligation not to take it from you. However, the obligation is not unconditional: it depends in part upon the existence of a corresponding desire in you. If you do not care whether I take your car, then I generally do not violate your right by doing so.[9]

Tooley admits that it is difficult to formulate the connections between rights and desires precisely, because there are problem cases like people who are asleep or temporarily unconscious. He does not want to say that such people have no rights because they have, at that moment, no desires. Nevertheless, Tooley holds, the possession of a right must in some way be linked with the capacity to have the relevant desires, if not with having the actual desires themselves.

The next step is to apply this view about rights to the case of the right to life. To put the matter as simply as possible—more simply than Tooley himself does and no doubt *too* simply—if the right to life is the right to continue existing as a distinct entity, then the desire relevant to possessing a right to life is the desire to continue existing as a distinct entity. But only a being who is capable of conceiving of herself as a distinct entity existing over time—that is, only a person—could have this desire. Therefore only a person could have a right to life.

This is how Tooley first formulated his position, in a striking article entitled "Abortion and Infanticide," first published in 1972. The problem of how precisely to formulate the connections between rights and desires, however, led Tooley to alter his position in a subsequent book with the same title, *Abortion and Infanticide*. He there argues that an individual cannot at a given time—say, now—have a right to continued existence unless the individual is of a kind such that it can now be in its interests that it continue to exist. One might think that this makes a dramatic difference to the outcome of Tooley's position, for while a newborn infant would not seem to be capable of conceiving of itself as a distinct entity existing over time, we commonly think that it can be in the interests of an infant to be saved from death, even if the death would have been entirely without pain or suffering. We certainly do this in retrospect: I might say, if I know that I nearly died in infancy, that the person who snatched my pram from the path of the speeding train is my greatest benefactor, for without her swift thinking I would never have had the happy and fulfilling life that I am now living. Tooley argues, however, that the retrospective attribution of an interest in living to the infant is a mistake. I am not the infant from whom I developed. The infant could not look forward to developing into the kind of being I am, or even into any intermediate being, between the being I now am and the infant. I cannot even recall being the infant; there are no mental links between us. Continued existence cannot be in the interests of a being who *never* has had the concept of a continuing self—that is, never has been able to conceive of itself as existing over time. If the train had instantly killed the infant, the death would not have been contrary to the interests of the infant, because the infant would never have had the concept of existing over time. It is true that I would then not be alive, but I can say that it is in my interests to be alive only because I do have the concept of a continuing self. I can with equal truth say that it is in my interests that my parents met, because if they had never met, they could not have created the embryo from which I developed, and so I would not be alive. This does not mean that the creation of this embryo was in the interests of any potential being who was lurking around, waiting to be brought into existence. There was no such being, and had I not been brought into existence, there would not have been anyone who missed out on the life I have enjoyed living. Similarly, we make a mistake if we now construct an interest in future life in the infant, who in the first days following birth can have no concept of continued existence, and with whom I have no mental links.

Hence in his book Tooley reaches, though by a more circuitous route, a conclusion that is practically equivalent to the conclusion he reached in his article. To have a right to life, one must have, or at least at one time have had, the concept of having a continued existence. Note that this formulation avoids any problems in dealing with sleeping or unconscious people; it is enough that they have had, at one time, the concept of continued existence for us to be able to say that continued life may be in their interests. This makes sense: my desire to continue living—or to complete the book I am writing, or to travel around the world next year—does not cease whenever I am not consciously thinking about these things. We often desire things without having the desire at the forefront of our minds. The fact that we have the desire is apparent if we are reminded of it, or suddenly confronted with a situation in which we must choose between two courses of action, one of which makes the fulfilment of the desire less likely. In a similar way, when we go to sleep our desires for the future have not ceased to exist. They will still be there when we wake. As the desires are still part of us, so, too, our interest in continued life remains part of us while we are asleep or unconscious.

People and Respect for Autonomy

To this point our discussion of the wrongness of killing people has focused on their capacity to envisage their future and have desires related to it. Another implication of being a person may also be relevant to the wrongness of killing. There is a strand of ethical thought, associated with Kant but including many modern writers who are not Kantians, according to which respect for autonomy is a basic moral principle. By "autonomy" is meant the capacity to choose, to make and act on one's own decisions. Rational and self-conscious beings presumably have this ability, whereas beings who cannot consider the alternatives open to them are not capable of choosing in the required sense and hence cannot be autonomous. In particular, only a being who can grasp the difference between dying and continuing to live can autonomously choose to live. Hence killing a person who does not choose to die fails to respect that person's autonomy; and as the choice of living or dying is about the most fundamental choice anyone can make, the choice on which all other choices depend, killing a person who does not choose to die is the gravest possible violation of that person's autonomy.

Not everyone agrees that respect for autonomy is a basic moral principle, or a valid moral principle at all. Utilitarians do not respect autonomy for its own sake, although they might give great weight to a person's desire to go on living, either in a preference utilitarian way or as evidence that the person's life was on the whole a happy one. But if we are preference utilitarians we must allow that a desire to go on living can be outweighed by other desires, and if we are classical utilitarians we must recognize that people may be utterly mistaken in their expectations of happiness. So a utilitarian, in objecting to the killing of a person, cannot place the same stress on autonomy as those who take respect for autonomy as an independent moral principle. The classical utilitarian might have to accept that in some cases it would be right to kill a person who does not choose to die on the grounds that the person will otherwise lead a miserable life. This is true, however, only on the critical level of moral reasoning. As we saw earlier, utilitarians may encourage people to adopt, in their daily lives, principles that will in almost all cases lead to better consequences when followed than any alternative action. The principle of respect for autonomy would be a prime example of such a principle.[10]

It may be helpful here to draw together our conclusions about the value of a person's life. We have seen that there are four possible reasons for holding that a person's life has some distinctive value over and above the life of a merely sentient being: the classical utilitarian concern with the effects of the killing on others; the preference utilitarian concern with the frustration of the victim's desires and plans for the future; the argument that the capacity to conceive of oneself as existing over time is a necessary condition of a right to life; and respect for autonomy. Although at the level of critical reasoning a classical utilitarian would accept only the first, indirect, reason, and a preference utilitarian only the first two reasons, at the intuitive level utilitarians of both kinds would probably advocate respect for autonomy too. The distinction between critical and intuitive levels thus leads to a greater degree of convergence, at the level of everyday moral decision making, between utilitarians and those who hold other moral views than we would find if we took into account only the critical level of reasoning. In any case, none of the four reasons for giving special protection to the lives of persons can be rejected out of hand. We shall therefore bear all four in mind when we turn to practical issues involving killing.

Before we do turn to practical questions about killing, however, we have still to consider claims about the value of life that are based neither on membership in our species nor on being a person.

Conscious Life

THERE ARE MANY BEINGS who are sentient and capable of experiencing pleasure and pain but are not rational and self-conscious and so not persons. I shall refer to these as conscious beings. Many nonhuman animals almost certainly fall into this category; so must newborn infants and some intellectually disabled humans. Exactly which of these lack self-consciousness is something we shall consider in the next chapters. If Tooley is right, those beings who do lack self-consciousness cannot be said to have a right to life, in the full sense of "right." Still, for other reasons, it might be wrong to kill them. In the present section we shall ask if the life of a being who is conscious but not self-conscious has value, and if so, how the value of such a life compares with the value of a person's life.

Should We Value Conscious Life?

The most obvious reason for valuing the life of a being capable of experiencing pleasure or pain is the pleasure it can experience. If we value our own pleasures—like the pleasures of eating, of sex, of running at full speed, and of swimming on a hot day—then the universal aspect of ethical judgments requires us to extend our positive evaluation of our own experience of these pleasures to the similar experiences of all who can experience them. But death is the end of all pleasurable experiences. Thus the fact that beings will experience pleasure in the future is a reason for saying that it would be wrong to kill them. Of course, a similar argument about pain points in the opposite direction, and it is only when we believe that the pleasure that beings are likely to experience outweighs the pain they are likely to suffer that this argument counts against killing. So what this amounts to is that we should not cut short a pleasant life.

This seems simple enough: we value pleasure; killing those who lead pleasant lives eliminates the pleasure they would otherwise experience; therefore such killing is wrong. But stating the argument in this way conceals something that, once noticed, makes the issue anything but simple. There are two ways of reducing the amount of pleasure in the world: one is to eliminate pleasures from the lives of those leading pleasant lives; the other is to eliminate those leading pleasant lives. The former leaves behind beings who experience less pleasure than they otherwise would have. The latter does not. This means that we cannot move automatically from a preference for a pleasant life rather than an unpleasant one to a preference for a pleasant life rather than no life at all. For, it might be objected, being killed does not make us worse off; it makes us cease to exist. Once we have ceased to exist, we shall not miss the pleasure we would have experienced.

Perhaps this seems sophistical—an instance of the ability of academic philosophers to find distinctions where there are no significant differences. If that is what you think, consider the opposite case: a case not of reducing pleasure but of increasing it. There are two ways of increasing the amount of pleasure in the world: one is to increase the pleasure of those who now exist; the other is to increase the number of those who will lead pleasant lives. If killing those leading pleasant lives is bad because of the loss of pleasure, then it would seem to be good to increase the number of those leading pleasant lives. We could do this by having more children, provided we could reasonably expect their lives to be pleasant, or by rearing large numbers of animals under conditions that would ensure that their lives would be pleasant. But would it really be good to create more pleasure by creating more pleased beings?

There seem to be two possible approaches to these perplexing issues. The first approach is simply to accept that it is good to increase the amount of pleasure in the world by increasing the number of pleasant lives, and bad to reduce the amount of pleasure in the world by reducing the number of pleasant lives. This approach has the advantage of being straightforward and clearly consistent, but it requires us to hold that if we could increase the number of beings leading pleasant lives without making others worse off, it would be good to do so. To see whether you are troubled by this conclusion, it may be helpful to consider a specific case. Imagine that a couple are trying to decide whether to have children. Suppose that as far as their own happiness is concerned, the advantages and disadvantages balance out. Children will interfere with their careers at a crucial stage of their professional lives and they will have to give up their favorite recreation, cross-country skiing, for a few years at least. At the same time, they know that, like most parents, they will get joy and fulfillment from having children and watching them develop. Suppose that if others will be affected, the good and bad effects will cancel each other out. Finally, suppose that since the couple could provide their children with a good start in life, and the children would be citizens of a developed nation with a high living standard, it is probable that their children will lead pleasant lives. Should the couple count the likely future pleasure of their children as a significant reason for having children? I doubt that many couples would, but if we accept this first approach, they should.

I shall call this approach the "total" view, since on this view we aim to increase the total amount of pleasure (and reduce the total amount of pain) and are indifferent whether this is done by increasing the pleasure of existing beings or increasing the number of beings who exist.

The second approach is to count only beings who already exist, prior to the decision we are taking, or at least will exist independently of that decision. We can call this the "prior existence" view. It denies that there is value in increasing pleasure by creating additional beings. The prior existence view is more in harmony with the intuitive judgment most people have (I think) that couples are under no moral obligation to have children when the children are likely to lead pleasant lives and no one else is adversely affected. But how do we square the prior existence view with our intuitions about the reverse case, when a couple are considering having a child who, perhaps because it will inherit a genetic defect, would lead a thoroughly miserable life and die before its second birthday? We would think it wrong for a couple knowingly to conceive such a child; but if the pleasure a possible child will experience is not a reason for bringing it into the world, why is the pain a possible child will experience a reason *against* bringing it into the world? The prior existence view must either hold that there is nothing wrong with bringing a miserable being into the world, or explain the asymmetry between cases of possible children who are likely to have pleasant lives and possible children who are likely to have miserable lives. Denying that it is bad knowingly to bring a miserable child into the world is hardly likely to appeal to those who adopted the prior existence view in the first place because it seemed more in harmony with their intuitive judgments than the total view; but a convincing explanation of the asymmetry is not easy to find. Perhaps the best one can say—and it is not very good—is that there is nothing directly wrong in conceiving a child who will be miserable, but once such a child exists, since its life can contain nothing but misery, we should reduce the amount of pain in the world by an act of euthanasia. But euthanasia is a more harrowing process for the parents and others involved than nonconception. Hence we have an indirect reason for not conceiving a child bound to have a miserable existence.

So is it wrong to cut short a pleasant life? We can hold that it is, on either the total view or the prior existence view, but our answers commit us to different things in each case. We can take the prior existence approach only if we accept that it is not wrong to bring a miserable being into existence—or else offer an explanation for why this should be wrong, and yet it not be wrong to fail to bring into existence a being whose life will be pleasant. Alternatively we can take the total approach, but then we must accept that it is also good to create more beings whose lives will be pleasant—and this has some odd practical implications.[11]

Comparing the Value of Different Lives

If we can give an affirmative—albeit somewhat shaky—answer to the question whether the life of a being who is conscious but not self-conscious has some value, can we also compare the value of different lives, at different levels of consciousness or self-consciousness? We are not, of course, going to attempt to assign numerical values to the lives of different beings, or even to produce an ordered list. The best that we could hope for is some idea of the principles that, when supplemented with the appropriate detailed information about the lives of different beings, might serve as the basis for such a list. But the most fundamental issue is whether we can accept the idea of ordering the value of different lives at all.

Some say that it is anthropocentric, even speciesist, to order the value of different lives in a hierarchical manner. If we do so we shall, inevitably, be placing ourselves at the top and other beings closer to us in proportion to the resemblance between them and ourselves. Instead we should recognize that from the points of view of the different beings themselves, each life is of equal value. Those who take this view recognize, of course, that a person's life may include the study of philosophy while a mouse's life cannot; but they say that the pleasures of a mouse's life are all that the mouse has, and so can be presumed to mean as much to the mouse as the pleasures of a person's life mean to the person. We cannot say that the one is more or less valuable than the other.

Is it speciesist to judge that the life of a normal adult member of our species is more valuable than the life of a normal adult mouse? It would be possible to defend such a judgment only if we can find some neutral ground, some impartial standpoint from which we can make the comparison.

The difficulty of finding neutral ground is a very real practical difficulty, but I am not convinced that it presents an insoluble theoretical problem. I would frame the question we need to ask in the following manner. Imagine that I have the peculiar property of being able to turn myself into an animal, so that like Puck in *A Midsummer Night's Dream*,

"Sometimes a horse I'll be, sometimes a hound." And suppose that when I am a horse, I really am a horse, with all and only the mental experiences of a horse, and when I am a human being I have all and only the mental experiences of a human being. Now let us make the additional supposition that I can enter a third state in which I remember exactly what it was like to be a horse and exactly what it was like to be a human being. What would this third state be like? In some respects—the degree of self-awareness and rationality involved, for instance—it might be more like a human existence than an equine one, but it would not be a human existence in every respect. In this third state, then, I could compare horse-existence with human-existence. Suppose that I were offered the opportunity of another life, and given the choice of life as a horse or as a human being, the lives in question being in each case about as good as horse or human lives can reasonably be expected to be on this planet. I would then be deciding, in effect, between the value of the life of a horse (to the horse) and the value of the life of a human (to the human).

Undoubtedly this scenario requires us to suppose a lot of things that could never happen, and some things that strain our imagination. The coherence of an existence in which one is neither a horse nor a human, but remembers what it was like to be both, might be questioned. Nevertheless I think I can make some sense of the idea of choosing from this position; and I am fairly confident that from this position, some forms of life would be seen as preferable to others.

If it is true that we can make sense of the choice between existence as a mouse and existence as a human, then—whichever way the choice would go—we can make sense of the idea that the life of one kind of animal possesses greater value than the life of another; and if this is so, then the claim that the life of every being has equal value is on very weak ground. We cannot defend this claim by saying that every being's life is all-important for it, since we have now accepted a comparison that takes a more objective—or at least intersubjective—stance and thus goes beyond the value of the life of a being considered solely from the point of view of that being.

So it would not necessarily be speciesist to rank the value of different lives in some hierarchical ordering. How we should go about doing this is another question, and I have nothing better to offer than the imaginative reconstruction of what it would be like to be a different kind of being. Some comparisons may be too difficult. We may have to say that we have not the slightest idea whether it would be better to be a fish or a snake;

but then, we do not very often find ourselves forced to choose between killing a fish or a snake. Other comparisons might not be so difficult. In general it does seem that the more highly developed the conscious life of the being, the greater the degree of self-awareness and rationality and the broader the range of possible experiences, the more one would prefer that kind of life, if one were choosing between it and a being at a lower level of awareness. Can utilitarians defend such a preference? In a famous passage John Stuart Mill attempted to do so:

> Few human creatures would consent to be changed into any of the lower animals, for a promise of the fullest allowance of a beast's pleasures; no intelligent human being would consent to be a fool, no instructed person would be an ignoramus, no person of feeling and conscience would be selfish and base, even though they should be persuaded that the fool, the dunce, or the rascal is better satisfied with his lot than they are with theirs. . . . It is better to be a human being dissatisfied than a pig satisfied; better to be Socrates dissatisfied than a fool satisfied. And if the fool, or the pig, are of a different opinion, it is because they only know their own side of the question. The other party to the comparison knows both sides.[12]

As many critics have pointed out, this argument is weak. Does Socrates really know what it is like to be a fool? Can he truly experience the joys of idle pleasure in simple things, untroubled by the desire to understand and improve the world? We may doubt it. But another significant aspect of this passage is less often noticed. Mill's argument for preferring the life of a human being to that of an animal (with which most modern readers would be quite comfortable) is exactly paralleled by his argument for preferring the life of an intelligent human being to that of fool. Given the context and the way in which the term "fool" was commonly used in his day, it seems likely that by this he means what we would now refer to as a person with an intellectual disability. With this further conclusion some modern readers will be distinctly uncomfortable; but as Mill's argument suggests, it is not easy to embrace the preference for the life of a human over that of a nonhuman, without at the same time endorsing a preference for the life of a normal human being over that of another human at an intellectual level similar to that of the nonhuman in the first comparison.

Mill's argument is difficult to reconcile with classical utilitarianism, because it just does not seem true that the more intelligent being necessarily has a greater capacity for happiness; and even if we were to accept that the capacity is greater, the fact that, as Mill acknowledges, this capacity is less often filled (the fool is satisfied, Socrates is not) would have to be taken into consideration. Would a preference utilitarian have a better prospect of defending the judgments Mill makes? That would depend on how we compare different preferences, held with differing degrees of awareness and self-consciousness. It does not seem impossible that we should find ways of ranking such different preferences, but at this stage the question remains open.

Taking Life: The Embryo and the Fetus

FROM *Practical Ethics*

The Conservative Position

THE CENTRAL ARGUMENT against abortion, put as a formal argument, would go something like this:

First premise: It is wrong to kill an innocent human being.
Second premise: A human fetus is an innocent human being.
Conclusion: Therefore it is wrong to kill a human fetus.

The usual liberal response is to deny the second premise of this argument. So it is on whether the fetus is a human being that the issue is joined, and the dispute about abortion is often taken to be a dispute about when a human life begins.

On this issue the conservative position is difficult to shake. The conservative points to the continuum between the fertilized egg and the child, and challenges the liberal to point to any stage in this gradual process that marks a morally significant dividing line. Unless there is such a line, the conservative says, we must either upgrade the status of the earliest embryo to that of the child, or downgrade the status of the child to that of the embryo; but no one wants to allow children to be dispatched on the request of their parents, and so the only tenable position is to grant the fetus the protection we now grant the child.

Is it true that there is no morally significant dividing line between fertilized egg and child? Those commonly suggested are: birth, viability, quickening, and the onset of consciousness. Let us consider these in turn.

Birth

Birth is the most visible possible dividing line, and the one that would suit liberals best. It coincides to some extent with our sympathies—we are less disturbed at the destruction of a fetus we have never seen than at the death of a being we can all see, hear, and cuddle. But is this enough to make birth the line that decides whether a being may or may not be killed? The conservative can plausibly reply that the fetus/baby is the same entity, whether inside or outside the womb, with the same human features (whether we can see them or not) and the same degree of awareness and capacity for feeling pain. A prematurely born infant may well be *less* developed in these respects than a fetus nearing the end of its normal term. It seems peculiar to hold that we may not kill the premature infant but may kill the more developed fetus. The location of a being—inside or outside the womb—should not make that much difference to the wrongness of killing it.

Viability

If birth does not mark a crucial moral distinction, should we push the line back to the time at which the fetus could survive outside the womb? This overcomes one objection to taking birth as the decisive point, for it treats the viable fetus on a par with the infant, born prematurely, at the same stage of development. Viability is where the United States Supreme Court drew the line in *Roe* v. *Wade*.[1] The Court held that the state has a legitimate interest in protecting potential life, and this interest becomes "compelling" at viability "because the fetus then presumably has the capability of meaningful life outside the mother's womb." Therefore statutes prohibiting abortion after viability would not, the Court said, be unconstitutional. But the judges who wrote the majority decision gave no indication why the mere capacity to exist outside the womb should make such a difference to the state's interest in protecting potential life. After all, if we talk, as the Court does, of *potential* human life, then the nonviable fetus is as much a potential adult human as the viable fetus. (I shall return to this issue of potentiality shortly; but it is a different issue from the conservative argument we are now discussing, which claims that the fetus is a human being, not just a potential human being.)

There is another important objection to making viability the cutoff point. The point at which the fetus can survive outside the mother's body varies according to the state of medical technology. Thirty years ago it was generally accepted that a baby born more than two months premature could not survive. Now a six-month fetus—three months premature—can often be pulled through, thanks to sophisticated medical techniques, and fetuses born after as little as five and a half months of gestation have survived. This threatens to undermine the Supreme Court's neat division of pregnancy into trimesters, with the boundary of viability lying between the second and third trimesters.

In light of these medical developments, do we say that a six-month-old fetus should not be aborted now but could have been aborted without wrongdoing thirty years ago? The same comparison can also be made, not between the present and the past, but between different places. A six-month-old fetus might have a fair chance of survival if born in a city where the latest medical techniques are used, but no chance at all if born in a remote village in Chad or New Guinea. Suppose that for some reason a woman, six months pregnant, was to fly from New York to a New Guinea village and that, once she had arrived in the village, there was no way she could return quickly to a city with modern medical facilities. Are we to say that it would have been wrong for her to have an abortion before she left New York, but now that she is in the village she may go ahead? The trip does not change the nature of the fetus, so why should it remove its claim to life?

The liberal might reply that the fact that the fetus is totally dependent on the mother for its survival means that it has no right to life independent of her wishes. In other cases, however, we do not hold that total dependence on another person means that that person may decide whether one lives or dies. A newborn baby is totally dependent on its mother, if it happens to be born in an isolated area in which there is no other lactating woman, nor the means for bottle feeding. An elderly woman may be totally dependent on the son looking after her, and a hiker who breaks her leg five days' walk from the nearest road may die if her companion does not bring help. We do not think that in these situations the mother may take the life of her baby, the son of his aged mother, or the hiker of her injured companion. So it is not plausible to suggest that the dependence of the nonviable fetus on its mother gives her the right to kill it; and if dependence does not justify making viability the dividing line, it is hard to see what does.

Quickening

If neither birth nor viability marks a morally significant distinction, there is less still to be said for a third candidate, quickening. Quickening is the time when the mother first feels the fetus move, and in traditional Catholic theology, this was thought to be the moment at which the fetus gained its soul. If we accepted that view, we might think quickening important, since the soul is, on the Christian view, what marks humans off from animals. But the idea that the soul enters the fetus at quickening is an outmoded piece of superstition, discarded now even by Catholic theologians. Putting aside these religious doctrines makes quickening insignificant. It is no more than the time when the fetus is first felt to move of its own accord; the fetus is alive before this moment, and ultrasound studies have shown that fetuses do in fact start moving as early as six weeks after fertilization, long before they can be felt to move. In any case, the capacity for physical motion—or the lack of it—has nothing to do with the seriousness of one's claim for continued life. We do not see the lack of such a capacity as negating the claims of paralyzed people to go on living.

Consciousness

Movement might be thought to be indirectly of moral significance, in so far as it is an indication of some form of awareness—and as we have already seen, consciousness, and the capacity to feel pleasure or pain, are of real moral significance. Despite this, neither side in the abortion debate has made much mention of the development of consciousness in the fetus. Those opposed to abortion may show films about the "silent scream" of the fetus when it is aborted, but the intention behind such films is merely to stir the emotions of the uncommitted. Opponents of abortion really want to uphold the right to life of the human being from conception, irrespective of whether it is conscious or not. For those in favor of abortion, to appeal to the absence of a capacity for consciousness has seemed a risky strategy. On the basis of the studies showing that movement takes place as early as six weeks after fertilization, coupled with other studies that have found some brain activity as early as the seventh week, it has been suggested that the fetus could be capable of feeling pain at this early stage of pregnancy. That possibility has made liberals very wary of appealing to the onset of consciousness as a point at which the fetus has a right to life. (For further discussion of fetal awareness, see "In Place of the Old Ethic.")

The liberal search for a morally crucial dividing line between the newborn baby and the fetus has failed to yield any event or stage of development that can bear the weight of separating those with a right to life from those who lack such a right, in a way that clearly shows fetuses to be in the latter category at the stage of development when most abortions take place. The conservative is on solid ground in insisting that the development from the embryo to the infant is a gradual process.

Some Liberal Arguments

Some liberals do not challenge the conservative claim that the fetus is an innocent human being but argue that abortion is nonetheless permissible. I shall consider three arguments for this view.

The Consequences of Restrictive Laws

The first argument is that laws prohibiting abortion do not stop abortions but merely drive them underground. Women who want to have abortions are often desperate. They will go to backyard abortionists or try folk remedies. Abortion performed by a qualified medical practitioner is as safe as any medical operation, but attempts to procure abortions by unqualified people often result in serious medical complications and sometimes death. Thus the effect of prohibiting abortion is not so much to reduce the number of abortions performed as to increase the difficulties and dangers for women with unwanted pregnancies.

This argument has been influential in gaining support for more liberal abortion laws. It was accepted by the Canadian Royal Commission on the Status of Women, which concluded that: "A law that has more bad effects than good ones is a bad law. . . . As long as it exists in its present form thousands of women will break it."

The main point to note about this argument is that it is an argument against laws prohibiting abortion, not an argument against the view that abortion is wrong. This is an important distinction, often overlooked in the abortion debate. The present argument well illustrates the distinction, because one could quite consistently accept it and advocate that the law should allow abortion on request, while at the same time deciding oneself—if one were pregnant—or counselling another who was pregnant, that it would be wrong to have an abortion. It is a mistake to assume that the law should always enforce morality. It may be that, as alleged in the case of abortion, attempts to enforce right conduct lead to consequences no one wants, and no decrease in wrongdoing; or it may be that, as is proposed by the next argument we shall consider, there is an area of private ethics with which the law ought not to interfere.

So this first argument is an argument about abortion law, not about the ethics of abortion. Even within those limits, however, it is open to challenge, for it fails to meet the conservative claim that abortion is the deliberate killing of an innocent human being, and in the same ethical category as murder. Those who take this view of abortion will not rest content with the assertion that restrictive abortion laws do no more than drive women to backyard abortionists. They will insist that this situation can be changed, and the law properly enforced. They may also suggest measures to make pregnancy easier to accept for those women who become pregnant against their wishes. This is a perfectly reasonable response, given the initial ethical judgment against abortion, and for this reason the first argument does not succeed in avoiding the ethical issue.

Not the Law's Business?

The second argument is again an argument about abortion laws rather than the ethics of abortion. It uses the view that, as the report of a British government committee inquiring into laws about homosexuality and prostitution put it: "There must remain a realm of private morality and immorality that is, in brief and crude terms, not the law's business."[2] This view is widely accepted among liberal thinkers and can be traced back to John Stuart Mill's *On Liberty*. The "one very simple principle" of this work is, in Mill's words:

> That the only purpose for which power can be rightfully exercised over any member of a civilised community, against his will, is to prevent harm to others. . . . He cannot rightfully be compelled to do or forbear because it will be better for him to do so, because it will make him happier, because in the opinions of others, to do so would be wise or even right.[3]

Mill's view is often and properly quoted in support of the repeal of laws that create "victimless crimes"—like laws prohibiting homosexual relations between consenting adults, the use of marijuana and other drugs, prostitution, gambling and so on. Abortion is often included in this list, for example by the criminologist Edwin Schur in his book *Crimes Without Victims.*[4] Those who consider abortion a victimless crime say that, while everyone is entitled to hold and act on his or her own view about the morality of abortion, no section of the community should try to force others to adhere to its own particular view. In a pluralist society, we should tolerate others with different moral views and leave the decision to have an abortion up to the woman concerned.

The fallacy involved in numbering abortion among the victimless crimes should be obvious. The dispute about abortion is, largely, a dispute about whether or not abortion does have a "victim." Opponents of abortion maintain that the victim of abortion is the fetus. Those not opposed to abortion may deny that the fetus counts as a victim in any serious way. They might, for instance, say that a being cannot be a victim unless it has interests that are violated, and the fetus has no interests. But however this dispute may go, one cannot simply ignore it on the grounds that people should not attempt to force others to follow their own moral views. My view that what Hitler did to the Jews is wrong is a moral view, and if there were any prospect of a revival of Nazism I would certainly do my best to force others not to act contrary to this view. Mill's principle is defensible only if it is restricted, as Mill restricted it, to acts that do not harm others. To use the principle as a means of avoiding the difficulties of resolving the ethical dispute over abortion is to take it for granted that abortion does not harm an "other"—which is precisely the point that needs to be proved before we can legitimately apply the principle to the case of abortion.

A Feminist Argument

The last of the three arguments that seek to justify abortion without denying that the fetus is an innocent human being is that a woman has a right to choose what happens to her own body. This argument became prominent with the rise of the women's liberation movement and has been elaborated by American philosophers sympathetic to feminism. An influential argument has been presented by Judith Jarvis Thomson by means of an ingenious analogy. Imagine, she says, that you wake up one morning and find yourself in a hospital bed, somehow connected to an unconscious man in an adjacent bed. You are told that this man is a famous violinist with kidney disease. The only way he can survive is for his circulatory system to be plugged into the system of someone else with the same blood type, and you are the only person whose blood is suitable. So a society of music lovers kidnapped you, had the connecting operation performed, and there you are. Since you are now in a reputable hospital you could, if you choose, order a doctor to disconnect you from the violinist; but the violinist will then certainly die. On the other hand, if you remain connected for only (only?) nine months, the violinist will have recovered and you can be unplugged without endangering him.[5]

Thomson believes that if you found yourself in this unexpected predicament you would not be morally required to allow the violinist to use your kidneys for nine months. It might be generous or kind of you to do so, but to say this is, Thomson claims, quite different from saying that you would be doing wrong if you did not do it.

Note that Thomson's conclusion does not depend on denying that the violinist is an innocent human being, with the same right to life as any other innocent human being. On the contrary, Thomson affirms that the violinist does have a right to life—but to have a right to life does not, she says, entail a right to the use of another's body, even if without that use one will die.

The parallel with pregnancy, especially pregnancy due to rape, should be obvious. A woman pregnant through rape finds herself, through no choice of her own, linked to a fetus in much the same way as the person is linked to the violinist. True, a pregnant woman does not normally have to spend nine months in bed, but opponents of abortion would not regard this as a sufficient justification for abortion. Giving up a newborn baby for adoption might be more difficult, psychologically, than parting from the violinist at the end of his illness; but this in itself does not seem a sufficient reason for killing the fetus. Accepting for the sake of the argument that the fetus does count as a full-fledged human being, having an abortion when the fetus is not viable has the same moral significance as unplugging oneself from the violinist. So if we agree with Thomson that it would not be wrong to unplug oneself from the violinist, we must also accept that, whatever the status of the fetus, abortion is not wrong—at least not when the pregnancy results from rape.

Thomson's argument can probably be extended beyond cases of rape. Suppose that you found yourself connected to the violinist, not because you were kidnapped by music lovers, but because you had intended to enter the hospital to visit a sick friend, and when you got into the elevator, you carelessly pressed the wrong button and ended up in a section of the hospital normally visited only by those who have volunteered to be connected to patients who would not otherwise survive. A team of doctors, waiting for the next volunteer, assumed you were it, jabbed you with an anesthetic, and connected you. If Thomson's argument was sound in the kidnapping case it is probably sound here too, since nine months of unwillingly supporting another is a high price to pay for ignorance or carelessness. In this way the argument might apply beyond rape cases to the much larger number of women who become pregnant through ignorance, carelessness, or contraceptive failure.

But is the argument sound? The short answer is this: It is sound if the particular theory of rights that lies behind it is sound, and it is unsound if that theory of rights is unsound.

The theory of rights in question can be illustrated by another of Thomson's fanciful examples: suppose I am desperately ill and the only thing that can save my life is the touch of my favorite film star's cool hand on my fevered brow. Well, Thomson says, even though I have a right to life, this does not mean that I have a right to force the film star to come to me, or that he is under any moral obligation to fly over and save me—although it would be frightfully nice of him to do so. Thus Thomson does not accept that we are always obliged to take the best course of action, all things considered, or to do what has the best consequences. She accepts, instead, a system of rights and obligations that allows us to justify our actions independently of their consequences.

A utilitarian would reject this theory of rights and would reject Thomson's judgment in the case of the violinist. The utilitarian would hold that, however outraged I may be at having been kidnapped, if the consequences of disconnecting myself from the violinist are, on balance, and taking into account the interests of everyone affected, worse than the consequences of remaining connected, I ought to remain connected. This does not necessarily mean that utilitarians would regard a woman who disconnected herself as wicked or deserving of blame. They might recognize that she has been placed in an extraordinarily difficult situation, one in which to do what is right involves a considerable sacrifice. They might even grant that most people in this situation would follow self-interest rather than do the right thing. Nevertheless, they would hold that to disconnect oneself is wrong.

In rejecting Thomson's theory of rights, and with it her judgment in the case of the violinist, the utilitarian would also be rejecting her argument for abortion. Thomson claimed that her argument justified abortion even if we allowed the life of the fetus to count as heavily as the life of a normal person. The utilitarian would say that it would be wrong to refuse to sustain a person's life for nine months, if that was the only way the person could survive. Therefore if the life of the fetus is given the same weight as the life of a normal person, the utilitarian would say that it would be wrong to refuse to carry the fetus until it can survive outside the womb.

This concludes our discussion of the usual liberal replies to the conservative argument against abortion. We have seen that liberals have failed to establish a morally significant dividing line between the newborn baby and the fetus, and their arguments—with the possible exception of Thomson's argument if her theory of rights can be defended—also fail to justify abortion in ways that do not challenge the conservative claim that the fetus is an innocent human being. Nevertheless, it would be premature for conservatives to assume that their case against abortion is sound. It is now time to bring into this debate some more general conclusions about the value of life.

The Value of Fetal Life

LET US GO BACK to the beginning. The central argument against abortion from which we started was:

First premise: It is wrong to kill an innocent human being.
Second premise: A human fetus is an innocent human being.
Conclusion: Therefore it is wrong to kill a human fetus.

The first set of replies we considered accepted the first premise of this argument but objected to the second. The second set of replies rejected neither premise, but objected to drawing the conclusion from these premises (or objected to the further conclusion that abortion should be prohibited by law). None of the replies questioned the first premise of the argument. Given the widespread acceptance of the doctrine of the sanctity of human life, this is not surprising; but the discussion of this doctrine in the preceding chapters shows that this premise is less secure than many people think.

The weakness of the first premise of the conservative argument is that it relies on our acceptance of the special status of *human* life. We have seen that "human" is a term that straddles two distinct notions: being a member of the species *Homo sapiens*, and being a person. Once the term is dissected in this way, the weakness of the conservative's first premise becomes apparent. If "human" is taken as equivalent to "person," the second premise of the argument, which asserts that the fetus is a human being, is clearly false; for one cannot plausibly argue that a fetus is either rational or self-conscious. If, on the other hand, "human" is taken to mean no more than "member of the species *Homo sapiens*," then the conservative defense of the life of the fetus is based on a characteristic lacking moral significance and so the first premise is false. The point should by now be familiar: whether a being is or is not a member of our species is, in itself, no more relevant to the wrongness of killing it than whether it is or is not a member of our race. The belief that mere membership in our species, irrespective of other characteristics, makes a great difference to the wrongness of killing a being is a legacy of religious doctrines that even those opposed to abortion hesitate to bring into the debate.

Recognizing this simple point transforms the abortion issue. We can now look at the fetus for what it is—the actual characteristics it possesses—and can value its life on the same scale as the lives of beings with similar characteristics who are not members of our species. It now becomes apparent that the "pro-life" or "right to life" movement is misnamed. Far from having concern for all life, or a scale of concern impartially based on the nature of the life in question, those who protest against abortion but dine regularly on the bodies of chickens, pigs, and calves show only a biased concern for the lives of members of our own species. For on any fair comparison of morally relevant characteristics, like rationality, self-consciousness, awareness, autonomy, pleasure, pain, and so on, the calf, the pig, and the much-derided chicken come out well ahead of the fetus at any stage of pregnancy—while if we make the comparison with a fetus of less than three months, a fish would show more signs of consciousness.

My suggestion, then, is that we accord the life of a fetus no greater value than the life of a nonhuman animal at a similar level of rationality, self-consciousness, awareness, capacity to feel, etc. Since no fetus is a person, no fetus has the same claim to life as a person. We have yet to consider at what point the fetus is likely to become capable of feeling pain. For now it will be enough to say that until that capacity exists, an abortion terminates an existence that is of no "intrinsic" value at all. Afterward, when the fetus may be conscious, though not self-conscious, abortion should not be taken lightly (if a woman ever does take abortion lightly).

But a woman's serious interests would normally override the rudimentary interests even of a conscious fetus. Indeed, even an abortion late in pregnancy for the most trivial reasons is hard to condemn unless we also condemn the slaughter of far more developed forms of life for the taste of their flesh.

The comparison between the fetus and other animals leads us to one more point. Where the balance of conflicting interests does make it necessary to kill a sentient creature, it is important that the killing be done as painlessly as possible. In the case of nonhuman animals the importance of humane killing is widely accepted; oddly, in the case of abortion little attention is paid to it. This is not because abortion is known to kill the fetus swiftly and humanely. Late abortions—which are the very ones in which the fetus may be able to suffer—are sometimes performed by injecting a salt solution into the amniotic sac that surrounds the fetus. It has been claimed that the effect of this is to cause the fetus to have convulsions and die between one and three hours later. Afterward the dead fetus is expelled from the womb. If there are grounds for thinking that a method of abortion causes the fetus to suffer, that method should be avoided.

The Fetus as Potential Life

ONE LIKELY OBJECTION to the argument I have offered in the preceding section is that it takes into account only the actual characteristics of the fetus, not its potential characteristics. On the basis of its actual characteristics, some opponents of abortion will admit, the fetus compares unfavourably with many nonhuman animals; it is when we consider its potential to become a mature human being that membership in the species *Homo sapiens* becomes important, and the fetus far surpasses any chicken, pig, or calf.

Up to this point I have not raised the question of the potential of the fetus because I thought it best to concentrate on the central argument against abortion; but it is true that a different argument, based on the potential of the fetus, can be mounted. Now is the time to look at this other argument. We can state it as follows:

First premise: It is wrong to kill a potential human being.
Second premise: A human fetus is a potential human being.
Conclusion: Therefore it is wrong to kill a human fetus.

The second premise of this argument is stronger than the second premise of the preceding argument. Whereas it is problematic whether a fetus actually *is* a human being—it depends on what we mean by the term—it cannot be denied that the fetus is a potential human being. This is true whether by "human being" we mean "member of the species *Homo sapiens*" or a rational and self-conscious being, a person. The strong second premise of the new argument is, however, purchased at the cost of a weaker first premise, for the wrongness of killing a potential human being—even a potential person—is more open to challenge than the wrongness of killing an actual human being.

It is of course true that the potential rationality, self-consciousness, and so on of a fetal *Homo sapiens* surpasses that of a cow or pig; but it does not follow that the fetus has a stronger claim to life. There is no rule that says that a potential X has the same value as an X or has all the rights of an X. There are many examples that show just the contrary. To pull out a sprouting acorn is not the same as cutting down a venerable oak. To drop a live chicken into a pot of boiling water would be much worse than doing the same to an egg. Prince Charles is a potential king of England, but he does not now have the rights of a king.

In the absence of any general inference from "A is a potential X" to "A has the rights of an X," we should not accept that a potential person should have the rights of a person, unless we can be given some specific reason why this should hold in this particular case. But what could that reason be? This question becomes especially pertinent if we recall the grounds on which, in the previous chapter, it was suggested that the life of a person merits greater protection than the life of a being who is not a person. These reasons—from the indirect classical utilitarian concern with not arousing in others the fear that they may be the next killed, the weight given by the preference utilitarian to a person's desires, Tooley's link between a right to life and the capacity to see oneself as a continuing mental subject, and the principle of respect for autonomy—are all based on the fact that persons see themselves as distinct entities with a past and future. They do not apply to those who are not now and never have been capable of seeing themselves in this way. If these are the grounds for not killing persons, the mere potential for becoming a person does not count against killing.

It might be said that this reply misunderstands the relevance of the potential of the human fetus, and that this potential is important, not because it creates in the fetus a right or claim to life, but because anyone who kills a human fetus deprives the world of a future rational and self-conscious being. If rational and self-conscious beings are intrinsically valuable, to kill a human fetus is to deprive the world of something intrinsically valuable, and so wrong. The chief problem with this as an argument against abortion—apart from the difficulty of establishing that rational and self-conscious beings are of intrinsic value—is that it does not stand up as a reason for objecting to all abortions, or even to abortions carried out merely because the pregnancy is inconveniently timed. Moreover the argument leads us to condemn practices other than abortion that most antiabortionists accept.

The claim that rational and self-conscious beings are intrinsically valuable is not a reason for objecting to all abortions because not every abortion deprives the world of a rational and self-conscious being. Suppose a woman has been planning to join a mountain-climbing expedition in June, and in January she learns that she is two months pregnant. She has no children at present and firmly intends to have a child within a year or two. The pregnancy is unwanted only because it is inconveniently timed. Opponents of abortion would presumably think an abortion in these circumstances particularly outrageous, for neither the life nor the health of the mother is at stake—only the enjoyment she gets from climbing mountains. Yet if abortion is wrong only because it deprives the world of a future person, this abortion is not wrong; it does no more than delay the entry of a person into the world.

On the other hand this argument against abortion does lead us to condemn practices that reduce the future human population: contraception, whether by "artificial" means or by "natural" means such as abstinence on days when the woman is likely to be fertile; and also celibacy. This argument has, in fact, all the difficulties of the "total" form of utilitarianism, discussed in the previous chapter, and it does not provide any reason for thinking abortion worse than any other means of population control. If the world is already overpopulated, the argument provides no reason at all against abortion.

Is there any other significance in the fact that the fetus is a potential person? If there is, I have no idea what it could be. In writings against abortion we often find reference to the fact that each human fetus is unique. Paul Ramsey, a former professor of religion at Princeton University, has said that modern genetics, by teaching us that the first fusion of sperm and ovum creates a "never-to-be-repeated" informational speck, seems to lead us to the conclusion that "all destruction of fetal life should be classified as murder."[6] But why should this fact lead us to this conclusion? A canine fetus is also, no doubt, genetically unique. Does this mean that it is as wrong to abort a dog as a human? When identical twins are conceived, the genetic information is repeated. Would Ramsey therefore think it permissible to abort one of a pair of identical twins? The children that my wife and I would produce if we did not use contraceptives would be genetically unique. Does the fact that it is still indeterminate precisely what genetically unique character those children would have make the use of contraceptives less evil than abortion? Why should it? And if it does, could the looming prospect of successful cloning—a technique in which the cells of one individual are used to reproduce a fetus that is a genetic carbon copy of the original—diminish the seriousness of abortion? Suppose the woman who wants to go mountain climbing were able to have her abortion, take a cell from the aborted fetus, and then reimplant that cell in her womb so that an exact genetic replica of the aborted fetus would develop—the only difference being that the pregnancy would now come to term six months later, and thus she could still join the expedition. Would that make the abortion acceptable? I doubt that many opponents of abortion would think so.

Abortion and Infanticide

THERE REMAINS one major objection to the argument I have advanced in favor of abortion. We have already seen that the strength of the conservative position lies in the difficulty liberals have in pointing to a morally significant line of demarcation between an embryo and a newborn baby. The standard liberal position needs to be able to point to some such line, because liberals usually hold that it is permissible to kill an embryo or fetus but not a baby. I have argued that the life of a fetus (and even more plainly, of an embryo) is of no greater value than the life of a nonhuman animal at a similar level of rationality, self-consciousness, awareness, capacity to feel, etc., and that since no fetus is a person no fetus has the same claim to life as a person. Now it must be admitted that these arguments apply to the newborn baby as much as to the fetus. A week-old baby is not a rational and self-conscious being, and there are many nonhuman animals whose rationality, self-consciousness, awareness, capacity to feel, and so on, exceed that of a human baby a week or a month old. If the fetus does not have the same claim to life as a person, it appears that the newborn baby does not either, and the life of a newborn baby is of less value to it than the life of a pig, a dog, or a chimpanzee is to the nonhuman animal. Thus while my position on the status of fetal life may be acceptable to many, the implications of this position for the status of newborn life are at odds with the virtually unchallenged assumption that the life of a newborn baby is as sacrosanct as that of an adult. Indeed, some people seem to think that the life of a baby is more precious than that of an adult. Lurid tales of German soldiers bayoneting Belgian babies figured prominently in the wave of anti-German propaganda that accompanied Britain's entry into the First World War, and it seemed to be tacitly assumed that this was a greater atrocity than the murder of adults would be.

I do not regard the conflict between the position I have taken and widely accepted views about the sanctity of infant life as a ground for abandoning my position. These widely accepted views need to be challenged. It is true that infants appeal to us because they are small and helpless, and there are no doubt very good evolutionary reasons why we should instinctively feel protective toward them. It is also true that infants cannot be combatants and killing infants in wartime is the clearest possible case of killing civilians, which is prohibited by international convention. In general, since infants are harmless and morally incapable of committing a crime, those who kill them lack the excuses often offered for the killing of adults. None of this shows, however, that the killing of an infant is as bad as the killing of an (innocent) adult.

In thinking about this matter we should put aside feelings based on the small, helpless, and—sometimes—cute appearance of human infants. To think that the lives of infants are of special value because infants are small and cute is on a par with thinking that a baby seal, with its soft white fur coat and large round eyes, deserves greater protection than a gorilla, who lacks these attributes. Nor can the helplessness or the innocence of the infant *Homo sapiens* be a ground for preferring it to the equally helpless and innocent fetal *Homo sapiens*, or, for that matter, to laboratory rats who are "innocent" in exactly the same sense as the human infant, and, in view of the experimenters' power over them, almost as helpless.

If we can put aside these emotionally moving but strictly irrelevant aspects of the killing of a baby, we can see that the grounds for not killing persons do not apply to newborn infants. The indirect, classical utilitarian reason does not apply, because no one capable of understanding what is happening when a newborn baby is killed could feel threatened by a policy that gave less protection to the newborn than to adults. In this respect Bentham was right to describe infanticide as "of a nature not to give the slightest inquietude to the most timid imagination."[7] Once we are old enough to comprehend the policy, we are too old to be threatened by it.

Similarly, the preference utilitarian reason for respecting the life of a person cannot apply to a newborn baby. Newborn babies cannot see themselves as beings who might or might not have a future, and so cannot have a desire to continue living. For the same reason, if a right to life must be based on the capacity to want to go on living, or on the ability to see oneself as a continuing mental subject, a newborn baby cannot have a right to life. Finally, a newborn baby is not an autonomous being, capable of making choices, and so to kill a newborn baby cannot violate the principle of respect for autonomy. In all this the newborn baby is on the same footing as the fetus, and hence fewer reasons exist against killing both babies and fetuses than exist against killing those who are capable of seeing themselves as distinct entities, existing over time.

It would, of course, be difficult to say at what age children begin to see themselves as distinct entities existing over time.[8] Even when we talk with two- and three-year-old children, it is usually very difficult to elicit any coherent conception of death, or of the possibility that someone—let alone the child herself—might cease to exist. No doubt children vary greatly in the age at which they begin to understand these matters, as they do in most things. But a difficulty in drawing the line is not a reason for drawing it in a place that is obviously wrong, any more than the notorious difficulty in saying how much hair a man has to have lost before we can call him "bald" is a reason for saying that someone whose pate is as smooth as a billiard ball is not bald. Of course, where rights are at risk, we should err on the side of safety. There is some plausibility in the view that, for legal purposes, since birth provides the only sharp, clear, and easily understood line, the law of homicide should continue to apply immediately after birth. Since this is an argument at the level of public policy and the law, it is quite compatible with the view that, on purely ethical grounds, the killing of a newborn infant is not comparable to the killing of an older child or adult. Alternatively, recalling Hare's distinction between the critical and intuitive levels of moral reasoning, one could hold that the ethical judgment we have reached applies only at the level of critical morality; for everyday decision-making, we should act as if an infant has a right to life from the moment of birth. It is, however, worth considering another possibility: that there should be at least some circumstances in which a full legal right to life comes into force not at birth, but only a short time after birth—perhaps a month. This would provide the ample safety margin mentioned above.

If these conclusions seem too shocking to take seriously, it may be worth remembering that our present absolute protection of the lives of infants is a distinctively Christian attitude rather than a universal ethical value. Infanticide has been practised in societies ranging geographically from Tahiti to Greenland and varying in culture from the nomadic Australian aborigines to the sophisticated urban communities of ancient Greece or mandarin China.[9] In some of these societies infanticide was not merely permitted but, in certain circumstances, deemed morally obligatory. Not to kill a deformed or sickly infant was often regarded as wrong, and infanticide was probably the first, and in several societies the only, form of population control.

We might think that we are just more "civilized" than these "primitive" peoples. But it is not easy to feel confident that we are more civilized than the best Greek and Roman moralists. It was not just the Spartans who exposed their infants on hillsides: both Plato and Aristotle recommended the killing of deformed infants.[10] Romans like Seneca, whose compassionate moral sense strikes the modern reader (or me, anyway) as superior to that of the early and medieval Christian writers, also thought infanticide the natural and humane solution to the problem posed by sick and deformed babies.[11] The change in Western attitudes toward infanticide since Roman times is, like the doctrine of the sanctity of human life of which it is a part, a product of Christianity. Perhaps it is now possible to think about these issues without assuming the Christian moral framework that has, for so long, prevented any fundamental reassessment.

None of this is meant to suggest that someone who goes around randomly killing babies is morally on a par with a woman who has an abortion. We should certainly put very strict conditions on permissible infanticide; but these restrictions might owe more to the effects of infanticide on others than to the intrinsic wrongness of killing an infant. Obviously, in most cases, to kill an infant is to inflict a terrible loss on those who love and cherish the child. My comparison of abortion and infanticide was prompted by the objection that the position I have taken on abortion also justifies infanticide. I have admitted this charge—without regarding the admission as fatal to my position—to the extent that the *intrinsic* wrongness of killing the late fetus and the *intrinsic* wrongness of killing the newborn infant are not markedly different. In cases of abortion, however, we assume that the people most affected—the parents-to-be, or at least the mother-to-be—want to have the abortion. Thus infanticide can be equated with abortion only when those closest to the child do not want it to live. As an infant can be adopted by others in a way that a pre-viable fetus cannot be, such cases will be rare. Killing an infant whose parents do not want it dead is, of course, an utterly different matter.

Prologue

FROM *Rethinking Life and Death*

AFTER RULING OUR THOUGHTS and our decisions about life and death for nearly two thousand years, the traditional western ethic has collapsed. To mark the precise moment when the old ethic gave way, a future historian might choose 4 February 1993, when Britain's highest court ruled that the doctors attending a young man named Anthony Bland could lawfully act to end the life of their patient. A Dutch historian, however, might choose instead 30 November 1993, the date on which the Netherlands parliament finally put into law the guidelines under which Dutch doctors had for some years already been openly giving lethal injections to patients who were suffering unbearably without hope of improvement, and who asked to be helped to die. Americans have not witnessed such momentous judicial decisions or votes in Congress, but twelve Michigan jurors may have spoken for the nation when, on 2 May 1994, they acquitted Dr. Jack Kevorkian of a charge of assisting a man named Thomas Hyde to commit suicide. Their refusal to convict Kevorkian was a major victory for the cause of physician-assisted suicide, for it is hard to imagine a clearer case than this one. Kevorkian freely admitted supplying the carbon monoxide gas, tubing, and a mask to Hyde, who had then used them to end a life made unbearable by a rapidly progressing nerve disorder known as Lou Gehrig's disease, or ALS.

These are the surface tremors resulting from major shifts deep in the bedrock of western ethics. We are going through a period of transition in our attitude toward the sanctity of human life. Such transitions cause confusion and division. That is why, for example, although the majority of people in most countries support laws allowing abortion, others sincerely believe that abortion is so great a wrong that they are prepared to block access to clinics that carry out abortions, to damage the clinics themselves, and even to go to the paradoxical extreme of murdering doctors because they performed abortions. Other symptoms of our bewilderment can be found everywhere. Here are three examples:

- The American Medical Association has a policy that says a doctor can ethically withdraw all means of life-prolonging medical treatment, *including food and water*, from a patient in an irreversible coma. Yet the same policy insists that "the physician should not intentionally cause death."[1]

- Twenty years after the introduction of "brain death" as a criterion of the death of a human being, one-third of doctors and nurses who work with brain-dead patients at hospitals in Cleveland, Ohio, thought that people whose brains had died could be classified as dead because they were "irreversibly dying" or because they had an "unacceptable quality of life."[2] Organ transplantation is based on the idea that we die when our brain is dead—yet even the doctors and nurses most closely involved do not really accept this.

- A recent survey asked pediatricians in senior positions in the United Kingdom to say whether they agreed or disagreed with a number of different statements, among which were these:

    1. Abortion is morally permissible after twenty-four weeks if the fetus is abnormal.
    2. There is no moral difference between the abortion of a fetus and the active termination of the life of a newborn infant when both have the same gestational age [that is, the same age dating from conception] and suffer from the same defects.
    3. There are no circumstances in which it is morally permissible to take active steps to terminate the life of an infant with severe defects.

Nearly 40 percent of the senior pediatricians responding indicated that they agreed with all three of these statements, even though you can't agree without contradicting yourself.

Each of these three examples is a snapshot that catches people halfway through a shift in their views. The American Medical Association has come to see the pointlessness of keeping people alive for ten, twenty, or thirty years if there is no hope that they will ever recover consciousness—but it has not yet summoned the nerve to abandon the traditional doctrine that it is always wrong to end the life of an innocent human being intentionally. Health-care professionals who work with organ transplantation have been taught that patients whose hearts still beat are dead because their brains are dead, but they have had difficulty reconciling this with their own feelings and way of thinking. Senior pediatricians have come to accept prenatal diagnosis and late termination of pregnancy if a serious abnormality is found. They also can see that there is no real difference between a late fetus and a newborn infant at the same gestational age. But active euthanasia for severely disabled infants remains illegal and is not sanctioned by medical codes of ethics, no matter how premature the infants may be, or how serious their defects.[3]

These are not academic problems found in the abstract theories of philosophers who remain remote from the real world, publishing papers in learned journals. These contradictions have direct consequences for human beings at the most deeply significant moments of their lives. The farce that the traditional ethic has become is also a tragedy that is endlessly repeated with minor variations in intensive care units all over the world. Since 1989, for me at least, the icon of this tragic farce has been the image of Rudy Linares, a twenty-three-year-old Chicago housepainter, standing in a hospital ward, keeping nurses at bay with a gun while he disconnects the respirator that for eight months has kept his comatose infant son Samuel alive. When Samuel is free of the respirator at last, Linares cradles him in his arms until, half an hour later, the child dies. Then Linares puts down the gun and, weeping, gives himself up. He acted against both the law and the traditional ethic that upholds the sanctity of human life; but his impulses were in accordance with an emerging ethical attitude that is more defensible than the old one, and will replace it.[4]

The traditional ethic is still defended by bishops and conservative bioethicists who speak in reverent tones about the intrinsic value of all human life, irrespective of its nature or quality. But, like the new clothes worn by the emperor, these solemn phrases seem true and substantial only while we are intimidated into uncritically accepting that all human life has some special dignity or worth. Once challenged, the traditional ethic crumples. Weakened by the decline in religious authority and the rise of a better understanding of the origins and nature of our species, that ethic is now being undone by changes in medical technology with which its inflexible strictures simply cannot cope.

This is not a cause for dismay or despair. A period of transition on so fundamental an issue is bound to be filled with uncertainty and confusion, especially among those who have been brought up to accept the traditional ethic as beyond question. But it is also a period of opportunity, in which we have a historic chance to shape something better, an ethic that does not need to be propped up by transparent fictions no one can really believe, an ethic that is more compassionate and more responsive to what people decide for themselves, an ethic that avoids prolonging life when to do so is obviously pointless, and an ethic that is less arbitrary in its inclusions and exclusions than our traditional one. To achieve a better approach to life-and-death decision-making, however, we need to be open about the ways in which the traditional ethic has failed.

Readers will already know that I do not speak in hushed tones when I refer to the traditional ethic of the sanctity of human life. Nor do I try to disguise its failings by invoking sophisticated distinctions and complex doctrines. I am not interested in continuing to patch and adjust the traditional approach so that we can pretend that it works when it plainly does not. The failures of the traditional ethic have become so glaring that these strategies can offer only short-term solutions to its problems, solutions that, like the policy of the American Medical Association on patients in irreversible coma, need to be reformulated almost as soon as they are pronounced. To break with the traditional way of approaching these issues is inevitably to clash with our usual moral beliefs. Some find this shocking. In part, this has more to do with the directness with which I describe what we already do than with any radically new suggestions I make about what we should do. When sensitive practices have long been veiled, ripping the veils aside can be shocking enough. But I readily admit that that is not the only reason why this book will shock readers. Some of the conclusions that I draw are very different from the ethical views most people hold today. That, however, is not a ground for dismissing them. If every proposal for reform in ethics that differed from accepted moral views had been rejected for that reason alone, we would still be torturing heretics, enslaving members of conquered races, and treating women as the property of their husbands. The views I put forward should be judged, not by the extent to which they clash with accepted moral views but on the basis of the arguments by which they are defended.

Supporters of the traditional sanctity of life ethic who know my previous writings will not be surprised to find themselves under siege in this book. But the book may also make uneasy some who have more affinity with the position I defend. This will include some who believe that they can coherently defend liberal abortion laws by saying that they are "pro-choice," without having to say when human life begins, or to show why the fetus should not count as a human being. Other natural allies of my position who may not like what I say here are those who have proposed changing the definition of brain death so that the death of the parts of the brain responsible for consciousness is sufficient for a patient to be declared dead. Even supporters of the present definition of brain death may be uncomfortable with the argument I put forward. These people may be disturbed because they have sought to present what they advocate in a form that leaves intact as much as possible of our traditional ethic. For those concerned with only one particular reform, that is a sensible political tactic—why take on the whole world when you can get by with only the antiabortion movement against you?

We can try to deal one at a time with the problems of the sanctity of life ethic. But the overall result will be a jigsaw puzzle, the pieces of which have to be forced into place, until the whole picture is under so much pressure that it buckles and breaks apart. I think there is a better way. There is a larger picture, in which all the pieces fit together. Whatever issue of the moment may concern us, in the long run we all need to see this larger picture. It will offer practical solutions to problems we now find insoluble, and allow us to act compassionately and humanely, where our ethic now leads us to outcomes that nobody wants.

Is the Sanctity of Life Ethic Terminally Ill?

FROM *Bioethics*

Revolution by Stealth: The Redefinition of Death

The acceptance of brain death—that is, the permanent loss of all brain function—as a criterion of death has been widely regarded as one of the great achievements of bioethics. It is one of the few issues on which there has been virtual consensus; and it has made an important difference in the way we treat people whose brains have ceased to function. This change in the definition of death has meant that warm, breathing, pulsating human beings are not given further medical support. If their relatives consent (or in some countries, as long as they have not registered a refusal of consent), their hearts and other organs can be cut out of their bodies and given to strangers. The change in our conception of death that excluded these human beings from the moral community was among the first in a series of dramatic changes in our view of life and death. Yet, in sharp contrast to other changes in this area, it met with virtually no opposition. How did this happen?

Everyone knows that the story of our modern definition of death begins with "The Ad Hoc Committee of the Harvard Medical School to Examine the Definition of Brain Death." What is not so well known is the link between the work of this committee and Dr. Christiaan Barnard's famous first transplantation of a human heart, in December 1967. Even before Barnard's sensational operation, Henry Beecher, chairman of a

Harvard University committee that oversaw the ethics of experimentation on human beings, had written to Robert Ebert, dean of the Harvard Medical School, suggesting that the committee should consider some new questions. He had, he told the dean, been speaking with Dr. Joseph Murray, who was a surgeon at Massachusetts General Hospital and a pioneer in kidney transplantation. "Both Dr. Murray and I," Beecher wrote, "think the time has come for a further consideration of the definition of death. Every major hospital has patients stacked up waiting for suitable donors."[1] Ebert did not respond immediately; but within a month of the news of the South African heart transplant, he set up, under Beecher's chairmanship, the group that was soon to become known as the Harvard Brain Death Committee.

The committee was made up mostly of members of the medical profession—ten of them, supplemented by a lawyer, a historian, and a theologian. It did its work rapidly and published its report in the *Journal of the American Medical Association* in August 1968. The report was soon recognized as an authoritative document, and its criteria for the determination of death were adopted rapidly and widely, not only in the United States but, with some modification of the technical details, in most countries of the world. The report began with a remarkably clear statement of what the committee was doing and why it needed to be done:

> Our primary purpose is to define irreversible coma as a new criterion for death. There are two reasons why there is a need for a definition: (1) Improvements in resuscitative and supportive measures have led to increased efforts to save those who are desperately injured. Sometimes these efforts have only a partial success so that the result is an individual whose heart continues to beat but whose brain is irreversibly damaged. The burden is great on patients who suffer permanent loss of intellect, on their families, on the hospitals, and on those in need of hospital beds already occupied by these comatose patients. (2) Obsolete criteria for the definition of death can lead to controversy in obtaining organs for transplantation.

To a reader familiar with bioethics in the 1990s, there are two striking aspects of this opening paragraph. The first is that the Harvard committee does not even attempt to argue that there is a need for a new definition of death because hospitals have a lot of patients in their wards who are really dead but are being kept attached to respirators because the law does not recognize them as dead. Instead, with unusual frankness, the committee said that a new definition was needed because irreversibly comatose patients were a great burden, not only on themselves (why to be in an irreversible coma is a burden to the patient, the committee did not say), but also to their families, hospitals, and patients waiting for beds. And then there was the problem of "controversy" about obtaining organs for transplantation.

In fact, frank as the statement seems, in presenting its concern about this controversy, the committee was still not being entirely candid. An earlier draft had been more open in stating that one reason for changing the definition of death was the "great need for tissues and organs of, among others, the patient whose cerebrum has been hopelessly destroyed, in order to restore those who are salvageable." When this draft was sent to Ebert, he advised Beecher to tone it down it because of its "unfortunate" connotation "that you wish to redefine death in order to make viable organs more readily available to persons requiring transplants."[2] The Harvard Brain Death Committee took Ebert's advice: it was doubtless more politic not to put things so bluntly. But Beecher himself made no secret of his own views. He was later to say, in an address to the American Association for the Advancement of Science:

> There is indeed a life-saving potential in the new definition, for, when accepted, it will lead to greater availability than formerly of essential organs in viable condition, for transplantation, and thus countless lives now inevitably lost will be saved. . . .[3]

The second striking aspect of the Harvard committee's report is that it keeps referring to "irreversible coma" as the condition that it wishes to define as death. The committee also speaks of "permanent loss of intellect" and even says, "We suggest that responsible medical opinion is ready to adopt new criteria for pronouncing death to have occurred in an individual sustaining irreversible coma as a result of permanent brain damage." Now "irreversible coma as a result of permanent brain damage" is by no means identical to the death of the whole brain. Permanent damage to the parts of the brain responsible for consciousness can also mean that a patient is in a "persistent vegetative state," a condition in which the brain stem and the central nervous system continue to function, but consciousness has been irreversibly lost. Even today, no legal system regards those in a persistent vegetative state as dead.

Admittedly, the Harvard committee report does go on to say, immediately following the paragraph quoted above: "*We are concerned here only with those comatose individuals who have no discernible central nervous system activity.*" But the reasons given by the committee for redefining death—the great burden on the patients, their families, the hospitals, and the community, as well as the waste of organs needed for transplation—apply in every respect to *all* those who are irreversibly comatose, not only to those whose entire brain is dead. So it is worth asking: why did the committee limit its concern to those with no brain activity at all? One reason could be that there was at the time no reliable way of telling whether a coma was irreversible, unless the brain damage was so severe that there was no brain activity at all. Another could be that people whose whole brain is dead will stop breathing after they are taken off a respirator and so will soon be dead by anyone's standard. People in a persistent vegetative state, on the other hand, may continue to breathe without mechanical assistance. To call for the undertakers to bury a "dead" patient who is still breathing would be a bit too much for anyone to swallow.

We all know that the redefinition of death proposed by the Harvard Brain Death Committee triumphed. By 1981, when the United States President's Commission for the Study of Ethical Problems in Medicine examined the issue, it could write of "the emergence of a medical consensus" around criteria very like those proposed by the Harvard committee.[4] Already, people whose brains had irreversibly ceased to function were considered legally dead in at least fifteen countries and in more than half of the states of the United States. In some countries, including Britain, the parliament had not even been involved in the change: the medical profession had simply adopted a new set of criteria on the basis of which doctors certified a patient dead.[5] This was truly a revolution without opposition.

The redefinition of death in terms of brain death went through so smoothly because it did not harm the brain-dead patients and it benefitted everyone else: the families of brain-dead patients, the hospitals, the transplant surgeons, people needing transplants, people who worried that they might one day need a transplant, people who feared that they might one day be kept on a respirator after their brain had died, taxpayers, and the government. The general public understood that if the brain has been destroyed, there can be no recovery of consciousness, and so there is no point in maintaining the body. Defining such people as dead was a convenient way around the problems of making their organs available for transplantation and withdrawing treatment from them.

But does this way around the problems really work? On one level, it does. By the early 1990s, as Sweden and Denmark, the last European nations to cling to the traditional standard, adopted brain-death definitions of death, this verdict appeared to be confirmed. Among developed nations, only Japan was still holding out.[6] But do people really think of the brain-dead as *dead*? The Harvard Brain Death Committee itself couldn't quite swallow the implications of what it was recommending. As we have seen, it described patients whose brains have ceased to function as in an "irreversible coma" and said that being kept on a respirator was a burden to them. Dead people are not in a coma; they are dead, and nothing can be a burden to them anymore.

Perhaps the lapses in the thinking of the Harvard committee can be pardoned because the concept of brain death was then so new. But twenty-five years later, little had changed. In 1993 the *Miami Herald* ran a story headlined "Brain-Dead Woman Kept Alive in Hopes She'll Bear Child"; while after the same woman did bear her child, the *San Francisco Chronicle* reported: "Brain-Dead Woman Gives Birth, Then Dies." Nor can we blame this entirely on the lamentable ignorance of the popular press. A study of doctors and nurses who work with brain-dead patients at hospitals in Cleveland, Ohio, showed that one in three of them thought that people whose brains had died could be classified as dead because they were "irreversibly dying" or because they had an "unacceptable quality of life."[7]

Why do both journalists and members of the health care professions talk in a way that denies that brain death is really death? One possible explanation is that even though people know that the brain-dead are dead, it is just too difficult for them to abandon obsolete ways of thinking about death. Another possible explanation is that people have enough common sense to see that the brain-dead are not really dead. I favor this second explanation. The brain death criterion of death is nothing other than a convenient fiction. It was proposed and accepted because it makes it possible for us to salvage organs that would otherwise be wasted, and to withdraw medical treatment when it is doing no good. On this basis, it might seem that, despite some fundamental weaknesses, the survival prospects of the concept of brain death are good. But our present understanding of brain death is not stable. Advances in medical knowledge and technology are the driving factors.

Brain death is generally defined as the irreversible cessation of all functions of the brain.[8] In accordance with this definition, a standard set of tests are used by doctors to establish that all functions of the brain have irreversibly ceased. These tests are broadly in line with those recommended in 1968 by the Harvard Brain Death Committee, but they have been further refined and updated over the years in various countries. In the past ten years, however, as doctors have sought ways of managing brain-dead patients so that their organs (or in some cases, their pregnancies) could be sustained for a longer time, it has become apparent that even when the usual tests show that brain death has occurred, *some brain functions continue.* We think of the brain primarily as concerned with processing information through the senses and the nervous system, but the brain has other functions as well. One of these is to supply various hormones that help to regulate several bodily functions. We now know that some of these hormones continue to be supplied by the brains of most patients who, by the standard tests, are brain-dead. Moreover, when brain-dead patients are cut open in order to remove organs, their blood pressure may rise and their heartbeat quicken. These reactions mean that the brain is still carrying out some of its functions, regulating the responses of the body in various ways. As a result, the legal definition of brain death and current medical practice in certifying brain-dead people as dead have come apart.[9]

It would be possible to bring medical practice into line with the current definition of death in terms of the irreversible cessation of *all* brain function. Doctors would then have to test for all brain functions, including hormonal functions, before declaring someone dead. This would mean that some people who are now declared brain-dead would be considered alive, and therefore would have to continue to be supported on a respirator, at significant cost, both financially and in terms of the extended distress of the family. Since the tests are expensive to carry out and time-consuming in themselves, continued support would be necessary during the period in which they are carried out, even if in the end the results showed that the person had no brain function at all. In addition, during this period, the person's organs would deteriorate and might therefore not be usable for transplantation. What gains would there be to balance against these serious disadvantages? From the perspective of an adherent of the sanctity of life ethic, of course, the gain is that we are no longer killing people by cutting out their hearts while they are still alive. If one really believed that the quality of a human life makes no difference to the wrongness of ending that life, this would end the discussion. There would be no ethical alternative. But it would still be true that not a single person who was kept longer on a respirator because of the need to test for hormonal brain functioning would ever return to consciousness.

So if it is life with consciousness, rather than life itself, that we value, then bringing medical practice into line with the definition of death does not seem a good idea. It would be better to bring the definition of brain death into line with current medical practice. But once we move away from the idea of brain death as the irreversible cessation of *all* brain functioning, what are we to put in its place? Which functions of the brain will we take as marking the difference between life and death, and why?

The most plausible answer is that the brain functions that really matter are those related to consciousness. On this view, what we really care about—and ought to care about—is the *person* rather than the body. Accordingly, it is the permanent cessation of function of the cerebral cortex, not of the whole brain, that should be taken as the criterion of death. Several reasons could be offered to justify this step. First, although the Harvard Brain Death Committee specified that its recommendations applied only to those who have "no discernible central nervous system activity," the arguments it put forward for its redefinition of death applied in every respect to patients who are permanently without any awareness, whether or not they have some brain stem function. This seems to have been no accident, for it reflected the view of the committee's chairman, Henry Beecher, who in his address to the American Association for the Advancement of Science, from which I have already quoted, said that what is essential to human nature is:

> . . . the individual's personality, his conscious life, his uniqueness, his capacity for remembering, judging, reasoning, acting, enjoying, worrying, and so on. . . .[10]

As I have already said, when the Harvard Committee issued its report, the irreversible destruction of the parts of the brain associated with consciousness could not reliably be diagnosed if the brain stem was alive. Since then, however, the technology for obtaining images of soft tissues within the body has made enormous progress. Hence a major stumbling block to the acceptance of a higher-brain definition of death has already been greatly diminished in its scope and will soon disappear altogether.

Now that medical certainty on the irreversibility of loss of higher brain functions can be established in at least some cases, the inherent logic of pushing the definition of death one step further has already led, in the United States, to one Supreme Court judge suggesting that the law could consider a person who has irreversibly lost consciousness to be no longer alive. Here is Mr. Justice Stevens, giving his judgment in the case of Nancy Cruzan, a woman who had been unconscious for eight years and whose guardians sought court permission to withdraw tube-feeding of food and fluids so that she could die:

> But for patients like Nancy Cruzan, who have no consciousness and no chance of recovery, there is a serious question as to whether the mere persistence of their bodies is "life," as that word is commonly understood. . . . The State's unflagging determination to perpetuate Nancy Cruzan's physical existence is comprehensible only as an effort to define life's meaning, not as an attempt to preserve its sanctity. . . . In any event, absent some theological abstraction, the idea of life is not conceived separately from the idea of a living person.[11]

Admittedly, this was a dissenting judgment; the majority decided the case on narrow constitutional grounds that are not relevant to our concerns here, and what Stevens said has not become part of the law of the United States. Nevertheless, dissenting judgments are often a way of floating an idea that is "in the air" and may become part of the majority view in a later decision. As medical opinion increasingly comes to accept that we can reliably establish when consciousness has been irreversibly lost, the pressure will become more intense for medical practice to move to a definition of death based on the death of the higher brain.

Yet there is a very fundamental flaw in the idea of moving to a higher-brain definition of death. If, as we have seen, people already have difficulty in accepting that a warm body with a beating heart on a respirator is really dead, how much more difficult would it be to bury a "corpse" that is still breathing while the lid of the coffin is nailed down? That is simply an absurdity. Something has gone wrong. But what?

In my view, the trouble began with the move to brain death. The Harvard Brain Death Committee was faced with two serious problems. Patients in an utterly hopeless condition were attached to respirators, and no one dared to turn them off; and organs that could be used to save lives were rendered useless by the delays caused by waiting for the circulation of the blood in potential donors to stop. The committee tried to solve both these problems by the bold expedient of classifying as dead those whose brains had ceased to have any discernible activity. The consequences of the redefinition of death were so evidently desirable that it met with scarcely any opposition and was accepted almost universally. Nevertheless, it was unsound from the start. Solving problems by redefinition rarely works, and this case was no exception. We need to begin again, with a different approach to the original problems, one which will break out of the intellectual straitjacket of the traditional belief that all human life is of equal value. Until 1993, it seemed difficult to imagine how a different approach could ever be accepted. But in that year Britain's highest court took a major step toward just such a new approach.

Revolution by the Law Lords: The Case of Anthony Bland

THE REVOLUTION in British law regarding the sanctity of human life grew out of a tragedy at Hillsborough Football Stadium in Sheffield, in April 1989. Liverpool was playing Nottingham Forest in a Football Association Cup semifinal. As the match started, thousands of supporters were still trying to get into the grounds. A fatal crush occurred against some fencing that had been erected to stop fans from getting onto the playing field. Before order could be restored and the pressure relieved, 95 people had died in the worst disaster in British sporting history. Tony Bland, a seventeen-year-old Liverpool fan, was not killed, but his lungs were crushed by the pressure of the crowd around him, and his brain was deprived of oxygen. When he was taken to the hospital, it was found that only his brain stem had survived. His cortex had been destroyed. Here is how Lord Justice Hoffmann was later to describe his condition:

> Since April 15 1989 Anthony Bland has been in persistent vegetative state. He lies in Airedale General Hospital in Keighley, fed liquid food by a pump through a tube passing through his nose and down the back of his throat into the stomach. His bladder is emptied through a catheter inserted through his penis, which from time to time has caused infections requiring dressing and antibiotic treatment. His stiffened joints have caused his limbs to be rigidly contracted so that his arms are tightly flexed across his chest and his legs unnaturally contorted. Reflex

> movements in the throat cause him to vomit and dribble. Of all this, and the presence of members of his family who take turns to visit him, Anthony Bland has no consciousness at all. The parts of his brain which provided him with consciousness have turned to fluid. The darkness and oblivion which descended at Hillsborough will never depart. His body is alive, but he has no life in the sense that even the most pitifully handicapped but conscious human being has a life. But the advances of modern medicine permit him to be kept in this state for years, even perhaps for decades.[12]

Whatever the advances of modern medicine might permit, neither Tony Bland's family nor his doctors could see any benefit to him or to anyone else in keeping him alive for decades. In Britain, as in many other countries, when everyone is in agreement in these situations it is quite common for the doctors simply to withdraw artificial feeding. The patient then dies within a week or two. In this case, however, the coroner in Sheffield was inquiring into the deaths caused by the Hillsborough disaster, and the doctor in charge of Tony Bland's care, Dr. Howe, decided that he should notify the coroner of what he was intending to do. The coroner, while agreeing that Bland's continued existence could well be seen as entirely pointless, warned Dr. Howe that he was running the risk of criminal charges—possibly even a charge of murder—if he intentionally ended Bland's life.

After the coroner's warning, the administrator of the hospital in which Bland was a patient applied to the Family Division of the High Court for declarations that the hospital might lawfully discontinue all life-sustaining treatment, including ventilation and the provision of food and water by artificial means, and discontinue all medical treatment to Bland "except for the sole purpose of enabling Anthony Bland to end his life and to die peacefully with the greatest dignity and the least distress."

At the Family Division hearing a public law officer called the Official Solicitor was appointed guardian for Bland for the purposes of the hearing. The Official Solicitor did not deny that Bland had no awareness at all and could never recover, but he nevertheless opposed what Dr. Howe was planning to do, arguing that, legally, it was murder. Sir Stephen Brown, president of the Family Division, did not accept this view, and he made the requested declarations to the effect that all treatment might lawfully be stopped. The Official Solicitor appealed, but Brown's decision was upheld by the court of appeal. The Official Solicitor then appealed again, thus bringing the case before the House of Lords.

We can best appreciate the significance of what the House of Lords did in the case of Tony Bland by looking at what the United States Supreme Court would not do in the similar case of Nancy Cruzan. Like Bland, Cruzan was in a persistent vegetative state, without hope of recovery. Her parents went to court to get permission to remove her feeding tube. The Missouri Supreme Court refused, saying that since Nancy Cruzan was not competent to refuse life-sustaining treatment herself, and the state has an interest in preserving life, the court could give permission for the withdrawal of life-sustaining treatment only if there was clear and convincing evidence that this was what Cruzan would have wanted. No such evidence had been presented to the court. On appeal, the United States Supreme Court upheld this judgment, ruling that the state of Missouri had a right to require clear and convincing evidence that Cruzan would have wanted to be allowed to die, before permitting doctors to take that step. (By a curious coincidence, that evidence was produced in court shortly after the Supreme Court decision, and Cruzan was allowed to die.)

The essential point here is that in America the courts have so far taken it for granted that life-support must be continued *unless* there is evidence indicating that the patient would not have wished to be kept alive in the circumstances in which she now is. In contrast, the British courts were quite untroubled by the absence of any information about what Bland's wishes might have been. As Sir Thomas Bingham, Master of the Rolls of the Court of Appeal, said in delivering his judgment:

> At no time before the disaster did Mr Bland give any indication of his wishes should he find himself in such a condition. It is not a topic most adolescents address.[13]

But the British courts did not therefore conclude that Bland must be treated until he died of old age. Instead, the British judges asked a different question: what is in the best interests of the patient?[14] In their answer, they referred to the unanimous medical opinion that Bland was not aware of anything and that there was no prospect of any improvement in his condition. Hence the treatment that was sustaining Bland's life brought him, as Sir Stephen Brown put in in the initial judgment in the case, "no therapeutical, medical, or other benefit."[15] In essence, the British courts held that when a patient is incapable of consenting to medical treatment, doctors are under no legal duty to continue treatment that does not benefit the patient. In addition, the judges agreed that the mere continuation of biological life is not, in the absence of any awareness or any hope of ever again becoming aware, a benefit to the patient.

On one level, the British approach is straightforward common sense. But it is common sense that breaks new legal ground. To see this, consider the following quotation from John Keown:

> Traditional medical ethics . . . never asks whether the patient's *life* is worthwhile, for the notion of a worthless life is as alien to the Hippocratic tradition as it is to English criminal law, both of which subscribe to the principle of the sanctity of human life which holds that, because all lives are intrinsically valuable, it is always wrong intentionally to kill an innocent human being.[16]

As a statement of traditional medical ethics and traditional English criminal law, this is right. The significance of the *Bland* decision is that it openly embraces the previously alien idea of a worthless life. Sir Thomas Bingham, for example, said:

> Looking at the matter as objectively as I can, and doing my best to look at the matter through Mr Bland's eyes and not my own, I cannot conceive what benefit his continued existence could be thought to give him. . . .[17]

When the case came before the House of Lords, their lordships took the same view. Lord Keith of Kinkel discussed the difficulties of making a value judgment about the life of a "permanently insensate" being, and concluded cautiously that:

> It is, however, perhaps permissible to say that to an individual with no cognitive capacity whatever, and no prospect of ever recovering any such capacity in this world, it must be a matter of complete indifference whether he lives or dies.[18]

In a similar vein, Lord Mustill concluded that to withdraw life support is not only legally but also ethically justified, "since the continued treatment of Anthony Bland can no longer serve to maintain that combination of manifold characteristics which we call a personality."[19]

There can therefore be no doubt that with the decision in the Bland case, British law has abandoned the idea that life itself is a benefit to the person living it, irrespective of its quality. But that is not all that their lordships did in deciding Tony Bland's fate. The second novel aspect of their decision is that it was as plain as anything can be that the proposal to discontinue tube feeding was *intended* to bring about Bland's death. A majority of the judges in the House of Lords referred to the administrator's intention in very direct terms. Lord Browne-Wilkinson said:

> What is proposed in the present case is to adopt a course with the intention of bringing about Anthony Bland's death. . . . The whole purpose of stopping artificial feeding is to bring about the death of Anthony Bland.[20]

Lord Mustill was equally explicit:

> The proposed conduct has the aim for . . . humane reasons of terminating the life of Anthony Bland by withholding from him the basic necessities of life.[21]

This marks a sharp contrast to what for many years was considered the definitive view of what a doctor may permissibly intend. Traditionally the law had held that while a doctor may knowingly do something that has the effect of shortening life, this must always be a mere side effect of an action with a different purpose, for example, relieving pain. As Justice (later Lord) Devlin said in the celebrated trial of Dr. John Bodkin Adams:

> . . . it remains the fact, and it remains the law, that no doctor, nor any man, no more in the case of the dying than of the healthy, has the right deliberately to cut the thread of human life.[22]

In rewriting the law of murder regarding the question of intention, the British law lords have shown a clarity and forthrightness that should serve as a model to many others who try to muddle through difficult questions by having a little bit of both sides. There is no talk here of ordinary and extraordinary means of treatment, nor of what is directly intended and what is merely foreseen. Instead the judges declared that Bland's doctors were entitled to take a course of action that had Bland's death as its "whole purpose"; and they made this declaration on the basis of a judgment that prolonging Bland's life did not benefit him.

Granted, this very clarity forces on us a further question: does the decision allow doctors to kill their patients? On the basis of what we have seen so far, this conclusion seems inescapable. Their lordships, however, did not think they were legalizing euthanasia. They drew a distinction between ending life by actively doing something, and ending life by not providing treatment needed to sustain life. That distinction has long been discussed by philosophers and bioethicists, who debate whether it can make good sense to accept passive euthanasia while rejecting active euthanasia. In the *Bland* case, it is significant that while the law lords insist that in distinguishing between acts and omissions they are merely applying the law as it stands, they explicitly recognize that at this point law and ethics have come apart, and something needs to be done about it. Lord Browne-Wilkinson, for example, expressed the hope that Parliament would review the law. He then ended his judgment by admitting that he could not provide a moral basis for the legal decision he had reached! Lord Mustill was just as frank and even more uncomfortable about the state of the law, saying that the judgment, in which he had shared, "may only emphasise the distortions of a legal structure which is already both morally and intellectually misshapen."[23]

The law lords' problem was that they had inherited a legal framework that allowed them some room to maneuver, but not a great deal. Within that framework, they did what they could to reach a sensible decision in the case of Anthony Bland, and to point the law in a new direction that other judges could follow. In doing so, they recognized the moral incoherence of the position they were taking but found themselves unable to do anything about it, beyond drawing the problem to the attention of Parliament. They could hardly have done more to show clearly the need for a new approach to life-and-death decisions.

Conclusion

WHAT IS THE LINK between the problems we face in regard to the concept of brain death, and the decision reached by their lordships in the case of Tony Bland? The link becomes clearer once we distinguish between three separate questions, often muddled in discussions of brain death and related issues:

1. When does a human being die?
2. When is it permissible for doctors to end the life of a patient intentionally?
3. When is it permissible to remove organs such as the heart from a human being for the purpose of transplantation to another human being?

Before 1968, in accordance with the traditional concept of death, the answer to the first question would have been: when the circulation of the blood stops permanently, with the consequent cessation of breathing, of a pulse, and so on.[24] The answer to the second question would then have been very simple: never. And the answer to the third question would have been equally plain: when the human being is dead.

The acceptance of the concept of brain death enabled us to hold constant the straightforward answers to questions 2 and 3, while making what was presented as no more than a scientific updating of a concept of death rendered obsolete by technological advances in medicine. Thus no ethical question appeared to be at issue, but suddenly hearts could be removed from, and machines turned off, a whole new group of human beings.

The *Bland* decision says nothing about questions 1 and 3, but it dramatically changes the answer that British law gives to question 2. The simple "never" now becomes "when the patient's continued life is of no benefit to her": and if we ask when a patient's life is of no benefit to her, the answer is: "when the patient is irreversibly unconscious." If we accept this as a sound answer to question 2, however, we may well wish to give the same answer to question 3. Why not, after all? And if we now have answered both question 2 and question 3 by reference, not to the death of the patient, but to the impossibility of the patient's regaining consciousness, then question 1 suddenly becomes much less relevant to the concerns that the Harvard Brain Death Committee was trying to address. We could therefore abandon the redefinition of death that it pioneered, with all the problems that have now arisen for the brain death criterion. Nor would we feel any pressure to move a step further, to defining death in terms of the death of the higher brain, or cerebral cortex. Instead, we could, without causing any problems in the procurement of organs or the withdrawal of life support, go back to the traditional conception of death in terms of the irreversible cessation of the circulation of the blood.

Justifying Infanticide

FROM *Practical Ethics*

IF WE WERE TO APPROACH the issue of life or death for a seriously disabled human infant without any prior discussion of the ethics of killing in general, we might be unable to resolve the conflict between the widely accepted obligation to protect the sanctity of human life and the goal of reducing suffering. Some say that such decisions are "subjective," or that life-and-death questions must be left to God and nature. Our previous discussions have, however, prepared the ground, and the principles established and applied in the preceding pages make the issue much less baffling than most take it to be.

The fact that a being is a human being, in the sense of a member of the species *Homo sapiens*, is not relevant to the wrongness of killing it; it is, rather, characteristics like rationality, autonomy, and self-consciousness that make a difference. Infants lack these characteristics. Killing them, therefore, cannot be equated with killing normal human beings, or any other self-conscious beings. This conclusion is not limited to infants who, because of irreversible intellectual disabilities, will never be rational, self-conscious beings. We saw in our discussion of abortion that the potential of a fetus to become a rational, self-conscious being cannot count against killing it at a stage when it lacks these characteristics—not, that is, unless we are also prepared to count the value of rational self-conscious life as a reason against contraception and celibacy. No infant—disabled or not—has as strong a claim to life as beings capable of seeing themselves as distinct entities, existing over time.

The difference between killing disabled and normal infants lies not in any supposed right to life that the latter has and the former lacks, but in other considerations about killing. Most obviously there is the difference that often exists in the attitudes of the parents. The birth of a child is usually a happy event for the parents. They have, nowadays, often planned for the child. The mother has carried it for nine months. From birth, a natural affection begins to bind the parents to it. So one important reason why it is normally a terrible thing to kill an infant is the effect the killing will have on its parents.

It is different when the infant is born with a serious disability. Birth abnormalities vary, of course. Some are trivial and have little effect on the child or its parents, but others turn the normally joyful event of birth into a threat to the happiness of the parents and any other children they may have.

Parents may, with good reason, regret that a disabled child was ever born. In that event the effect that the death of the child will have on its parents can be a reason for, rather than against killing it. Some parents want even the most gravely disabled infant to live as long as possible, and this desire would then be a reason against killing the infant. But what if this is not the case? In the discussion that follows I shall assume that the parents do not want the disabled child to live. I shall also assume that the disability is so serious that—again in contrast to the situation of an unwanted but normal child today—there are no other couples eager to adopt the infant. This is a realistic assumption even in a society in which there is a long waiting list of couples wishing to adopt normal babies. It is true that from time to time cases of infants who are severely disabled and are being allowed to die have reached the courts in a glare of publicity, and this has led to couples' offering to adopt the child. Unfortunately such offers are the product of the highly publicized dramatic life-and-death situation and do not extend to the less publicized but far more common situations in which parents feel themselves unable to look after a severely disabled child, and the child then languishes in an institution.

Infants are sentient beings who are neither rational nor self-conscious. So if we turn to consider the infants in themselves, independently of the attitudes of their parents, since their species is not relevant to their moral status, the principles that govern the wrongness of killing nonhuman animals who are sentient but not rational or self-conscious must apply here too. As we saw, the most plausible arguments for attributing a right to life to a being apply only if there is some awareness of oneself as a being existing over time, or as a continuing mental self. Nor can respect for autonomy apply where there is no capacity for autonomy. The remaining principles identified in the chapter "What's Wrong with Killing?" are utilitarian. Hence the quality of life that the infant can be expected to have is important.

One relatively common birth disability is a faulty development of the spine known as spina bifida. Its prevalence varies in different countries, but it can affect as many as one in five hundred live births. In the more severe cases, the child will be permanently paralyzed from the waist down and lack control of bowels or bladder. Often excess fluid accumulates in the brain, a condition known as hydrocephalus, which can result in intellectual disabilities. Though some forms of treatment exist, if the child is badly affected at birth, the paralysis, incontinence, and intellectual disability cannot be overcome.

Some doctors closely connected with children suffering from severe spina bifida believe that the lives of the worst affected children are so miserable that it is wrong to resort to surgery to keep them alive. Published descriptions of the lives of these children support the judgment that they will have lives filled with pain and discomfort. They need repeated major surgery to prevent curvature of the spine, due to the paralysis, and to correct other abnormalities. Some children with spina bifida have had forty major operations before they reach their teenage years.

When the life of an infant will be so miserable as not to be worth living, from the internal perspective of the being who will lead that life, both the "prior existence" and the "total" version of utilitarianism entail that, if there are no "extrinsic" reasons for keeping the infant alive—like the feelings of the parents—it is better that the child should be helped to die without further suffering. A more difficult problem arises—and the convergence between the two views ends—when we consider disabilities that make the child's life prospects significantly less promising than those of a normal child, but not so bleak as to make the child's life not worth living. Hemophilia is probably in this category. The hemophiliac lacks the element in normal blood that makes it clot and thus risks prolonged bleeding, especially internal bleeding, from the slightest injury. If allowed to continue, this bleeding leads to permanent crippling and eventually death. The bleeding is very painful, and although improved treatments have eliminated the need for constant blood transfusions, hemophiliacs still have to spend a lot of time in the hospital. They are unable to play most sports and live constantly on the edge of crisis. Nevertheless, hemophiliacs do not appear to spend their time wondering whether to end it all; most find life definitely worth living, despite the difficulties they face.

Given these facts, suppose that a newborn baby is diagnosed as a hemophiliac. The parents, daunted by the prospect of bringing up a child with this condition, are not anxious for him to live. Could euthanasia be defended here? Our first reaction may well be a firm "no," for the infant can be expected to have a life that is worth living, even if not quite as good as that of a normal baby. The "prior existence" version of utilitarianism supports this judgment. The infant exists. His life can be expected to contain a positive balance of happiness over misery. To kill him would deprive him of this positive balance of happiness. Therefore it would be wrong.

On the "total" version of utilitarianism, however, we cannot reach a decision on the basis of this information alone. The total view makes it necessary to ask whether the death of the hemophiliac infant would lead to the creation of another being who would not otherwise have existed. In other words, if the hemophiliac child is killed, will his parents have another child whom they would not have if the hemophiliac child lives? If they would, is the second child likely to have a better life than the one killed?

Often it will be possible to answer both these questions affirmatively. A woman may plan to have two children. If one dies while she is of childbearing age, she may conceive another in its place. Suppose a woman planning to have two children has one normal child and then gives birth to a hemophiliac child. The burden of caring for that child may make it impossible for her to cope with a third child; but if the disabled child were to die, she would have another. It is also plausible to suppose that the prospects of a happy life are better for a normal child than for a hemophiliac.

When the death of a disabled infant will lead to the birth of another infant with better prospects of a happy life, the total amount of happiness will be greater if the disabled infant is killed. The loss of happy life for the first infant is outweighed by the gain of a happier life for the second.

Therefore, if killing the hemophiliac infant has no adverse effect on others, it would, according to the total view, be right to kill him.

The total view treats infants as replaceable. Many will think that the replaceability argument cannot be applied to human infants. The direct killing of even the most hopelessly disabled infant is still officially regarded as murder; how then could the killing of infants with far less serious problems, like hemophilia, be accepted? Yet on further reflection, the implications of the replaceability argument do not seem quite so bizarre. For there are disabled members of our species whom we now deal with exactly as the argument suggests we should. These cases closely resemble the ones we have been discussing. There is only one difference, and that is a difference of timing—the timing of the discovery of the problem, and the consequent killing of the disabled being.

Prenatal diagnosis is now a routine procedure for pregnant women. There are various medical techniques for obtaining information about the fetus during the early months of pregnancy. At one stage in the development of these procedures, it was possible to discover the sex of the fetus, but not whether the fetus would suffer from hemophilia. Hemophilia is a sex-linked genetic defect, from which only males suffer; females can carry the gene and pass it on to their male offspring without themselves being affected. So a woman who knew that she carried the gene for hemophilia could, at that stage, avoid giving birth to a hemophiliac child only by finding out the sex of the fetus and aborting all male fetuses. Statistically, only half of these male children of women who carried the defective gene would have suffered from hemophilia, but there was then no way to find out to which half a particular fetus belonged. Therefore twice as many fetuses were being killed as necessary, in order to avoid the birth of children with hemophilia. This practice was widespread in many countries and yet did not cause any great outcry. Now that we have techniques for identifying hemophilia before birth, we can be more selective, but the principle is the same: women are offered, and usually accept, abortions in order to avoid giving birth to children with hemophilia.

The same can be said about some other conditions that can be detected before birth. Down syndrome, formerly known as mongolism, is one of these. Children with this condition have intellectual disabilities and most will never be able to live independently, but their lives, like those of small children, can be joyful. The risk of having a child with

Down syndrome increases sharply with the age of the mother, and for this reason prenatal diagnosis is routinely offered to pregnant women over 35. Again, undergoing the procedure implies that if the test for Down syndrome is positive, the woman will consider aborting the fetus and, if she still wishes to have another child, will start another pregnancy, which has a good chance of being normal.

Prenatal diagnosis, followed by abortion in selected cases, is common practice in countries with liberal abortion laws and advanced medical techniques. I think this is as it should be. As the arguments of the chapter "Taking Life: The Embryo and the Fetus" indicate, I believe that abortion can be justified. Note, however, that neither hemophilia nor Down syndrome is so crippling as to make life not worth living, from the inner perspective of the person with the condition. To abort a fetus with one of these disabilities, intending to have another child who will not be disabled, is to treat fetuses as interchangeable or replaceable. If the mother has previously decided to have a certain number of children, say two, then what she is doing, in effect, is rejecting one potential child in favor of another. She could, in defense of her actions, say: the loss of life of the aborted fetus is outweighed by the gain of a better life for the normal child who will be conceived only if the disabled one dies.

When death occurs before birth, replaceability does not conflict with generally accepted moral convictions. That a fetus is known to be disabled is widely accepted as a ground for abortion. Yet in discussing abortion, we saw that birth does not mark a morally significant dividing line. I cannot see how one could defend the view that fetuses may be "replaced" before birth, but newborn infants may not be. Nor is there any other point, such as viability, that does a better job of dividing the fetus from the infant. Self-consciousness, which could provide a basis for holding that it is wrong to kill one being and replace it with another, is not to be found in either the fetus or the newborn infant. Neither the fetus nor the newborn infant is an individual capable of regarding itself as a distinct entity with a life of its own to lead, and it is only for newborn infants, or for still earlier stages of human life, that replaceability should be considered to be an ethically acceptable option.

It may still be objected that to replace either a fetus or a newborn infant is wrong because it suggests to disabled people living today that their lives are less worth living than the lives of people who are not disabled. Yet it is surely flying in the face of reality to deny that, on average, this is so. That is the only way to make sense of actions that we all take for granted. Recall thalidomide: this drug, when taken by pregnant women, caused many children to be born without arms or legs. Once the cause of the abnormal births was discovered, the drug was taken off the market, and the company responsible had to pay compensation. If we really believed that there is no reason to think of the life of a disabled person as likely to be any worse than that of a normal person, we would not have regarded this as a tragedy. No compensation would have been sought, or awarded by the courts. The children would merely have been "different." We could even have left the drug on the market, so that women who found it a useful sleeping pill during pregnancy could continue to take it. If this sounds grotesque, that is only because we are all in no doubt at all that it is better to be born with limbs than without them. To believe this involves no disrespect at all for those who are lacking limbs; it simply recognizes the reality of the difficulties they face.

In any case, the position taken here does not imply that it would be better that no people born with severe disabilities should survive; it implies only that the parents of such infants should be able to make this decision. Nor does this imply lack of respect or equal consideration for people with disabilities who are now living their own lives in accordance with their own wishes. The principle of equal consideration of interests rejects any discounting of the interests of people on grounds of disability (see "All Animals Are Equal").

Even those who reject abortion and the idea that the fetus is replaceable are likely to regard possible people as replaceable. Suppose a woman is planning to become pregnant. She is told by her doctor that if she goes ahead with her plan to become pregnant immediately, her child will have hemophilia; but if she waits three months her child will be normal. If we think she would do wrong not to wait, it can only be because we are comparing the two possible lives and judging one to have better prospects than the other. Of course, at this stage no life has begun; but the question is, when does a life, in the morally significant sense, really begin? In earlier chapters we saw several reasons for saying that life begins in the morally significant sense only when there is awareness of one's existence over time.

Regarding newborn infants as replaceable, as we now regard fetuses, would have considerable advantages over prenatal diagnosis followed by abortion. Prenatal diagnosis still cannot detect all major disabilities. Some disabilities, in fact, are not present before birth; they may be the result of extremely premature birth or of something going wrong in the birth process itself. At present parents can choose to keep or destroy their disabled offspring only if the disability happens to be detected during pregnancy. There is no logical basis for restricting parents' choice to these particular disabilities. If disabled newborn infants were not regarded as having a right to life until, say, a week or a month after birth it would allow parents, in consultation with their doctors, to choose on the basis of far greater knowledge of the infant's condition than is possible before birth.

All these remarks have been concerned with the wrongness of ending the life of the infant, considered in itself rather than for its effects on others. When we take effects on others into account, the picture may alter. Obviously, to go through the whole of pregnancy and labor, only to give birth to a child who one decides should not live, would be a difficult, perhaps heartbreaking, experience. For this reason many women would prefer prenatal diagnosis and abortion rather than live birth with the possibility of infanticide; but if the latter is not morally worse than the former, this would seem to be a choice that the woman herself should be allowed to make.

Another factor to take into account is the possibility of adoption. When there are more couples wishing to adopt than normal children available for adoption, a childless couple may be prepared to adopt a hemophiliac. This would relieve the mother of the burden of bringing up a hemophiliac child and enable her to have another child, if she wished. Then the replaceability argument could not justify infanticide, for bringing the other child into existence would not be dependent on the death of the hemophiliac. The death of the hemophiliac would then be a straightforward loss of a life of positive quality, not outweighed by the creation of another being with a better life.

So the issue of ending life for disabled newborn infants is not without complications, which we do not have the space to discuss adequately. Nevertheless the main point is clear: killing a disabled infant is not morally equivalent to killing a person. Very often it is not wrong at all.

Justifying Voluntary Euthanasia

FROM *Practical Ethics*

UNDER EXISTING LAWS IN MOST COUNTRIES, people suffering unrelievable pain or distress from an incurable illness who beg their doctors to end their lives are asking their doctors to risk a murder charge. Although juries are extremely reluctant to convict in cases of this kind, the law is clear that neither the request, nor the degree of suffering, nor the incurable condition of the person killed, is a defense to a charge of murder. Advocates of voluntary euthanasia propose that this law be changed so that a doctor could legally act on a patient's desire to die without further suffering. Doctors have been able to do this quite openly in the Netherlands, as a result of a series of court decisions during the 1980s, as long as they comply with certain conditions. In Germany, doctors may provide a patient with the means to end her life, but they may not administer the substance to her.

The case for voluntary euthanasia has some common ground with the case for nonvoluntary euthanasia, in that death is a benefit for the one killed. The two kinds of euthanasia differ, however, in that voluntary euthanasia involves the killing of a person, a rational and self-conscious being and not a merely conscious being. (To be strictly accurate it must be said that this is not always so, because although only rational and self-conscious beings can consent to their own death, they may not be rational and self-conscious at the time euthanasia is contemplated—the doctor may, for instance, be acting on a prior written request for euthanasia if, through accident or illness, one's rational faculties should be irretrievably lost. For simplicity we shall, henceforth, disregard this complication.)

We have seen that it is possible to justify ending the life of a human being who lacks the capacity to consent. We must now ask in what way the ethical issues are different when the being is capable of consenting, and does in fact consent.

Let us return to the general principles about killing proposed in the chapter "What's Wrong with Killing?" I argued there that killing a self-conscious being is a more serious matter than killing a merely conscious being. I gave four distinct grounds on which this could be argued:

1. The classical utilitarian claim that since self-conscious beings are capable of fearing their own death, killing them has worse effects on others.

2. The preference utilitarian calculation that counts the thwarting of the victim's desire to go on living as an important reason against killing.

3. A theory of rights according to which to have a right one must have the ability to desire that to which one has a right, so that to have a right to life one must be able to desire one's own continued existence.

4. Respect for the autonomous decisions of rational agents.

Now suppose we have a situation in which a person suffering from a painful and incurable disease wishes to die. If the individual were not a person—not rational or self-conscious—euthanasia would, as I have said, be justifiable. Do any of the four grounds for holding that it is normally worse to kill a person provide reasons against killing when the individual is a person who wants to die?

The classical utilitarian objection does not apply to killing that takes place only with the genuine consent of the person killed. That people are killed under these conditions would have no tendency to spread fear or insecurity, since we have no cause to be fearful of being killed with our own genuine consent. If we do not wish to be killed, we simply do not consent. In fact, the argument from fear points in favor of voluntary euthanasia, for if voluntary euthanasia is not permitted we may, with good cause, be fearful that our deaths will be unnecessarily drawn out and distressing. In the Netherlands, a nationwide study commissioned by the government found that "Many patients want an assurance that their doctor will assist them to die should suffering become unbearable." Often, having received this assurance, the patients made no persistent request for euthanasia. The availability of euthanasia brought comfort without euthanasia having to be provided.[1]

Preference utilitarianism also points in favor of, not against, voluntary euthanasia. Just as preference utilitarianism must count a desire to go on living as a reason against killing, so it must count a desire to die as a reason for killing.

According to the theory of rights we have considered, it is an essential feature of a right that one can waive one's rights if one so chooses. I may have a right to privacy; but I can, if I wish, film every detail of my daily life and invite the neighbors to my home movies. Neighbors sufficiently intrigued to accept my invitation could do so without violating my right to privacy, since the right has on this occasion been waived. Similarly, to say that I have a right to life is not to say that it would be wrong for my doctor to end my life, if she does so at my request. In making this request I waive my right to life.

Last, the principle of respect for autonomy tells us to allow rational agents to live their own lives according to their own autonomous decisions, free from coercion or interference; but if rational agents should autonomously choose to die, then respect for autonomy will lead us to assist them to do as they choose.

So, although there are reasons for thinking that killing a self-conscious being is normally worse than killing any other kind of being, in the special case of voluntary euthanasia most of these reasons count for euthanasia rather than against it. Surprising as this result might at first seem, it really does no more than reflect the fact that what is special about self-conscious beings is that they can know that they exist over time and will, unless they die, continue to exist. Normally this continued existence is fervently desired; when the foreseeable continued existence is dreaded rather than desired however, the desire to die may take the place of the normal desire to live, reversing the reasons against killing based on the desire to live. Thus the case for voluntary euthanasia is arguably much stronger than the case for nonvoluntary euthanasia.

Some opponents of the legalisation of voluntary euthanasia might concede that all this follows, if we have a genuinely free and rational decision to die: but, they add, we can never be sure that a request to be killed is the result of a free and rational decision. Will not the sick and elderly be pressured by their relatives to end their lives quickly? Will it not be possible to commit outright murder by pretending that a person has requested euthanasia? And even if there is no pressure of falsification, can anyone who is ill, suffering pain, and very probably in a drugged and confused state of mind make a rational decision about whether to live or die?

These questions raise technical difficulties for the legalisation of voluntary euthanasia, rather than objections to the underlying ethical principles; but they are serious difficulties nonetheless. The guidelines developed by the courts in the Netherlands have sought to meet them by proposing that euthanasia is acceptable only if

- It is carried out by a physician.
- The patient has explicitly requested euthanasia in a manner that leaves no doubt of the patient's desire to die.
- The patient's decision is well-informed, free, and durable.
- The patient has an irreversible condition causing protracted physical or mental suffering that the patient finds unbearable.
- There is no reasonable alternative (reasonable from the patient's point of view) to alleviate the patient's suffering.
- The doctor has consulted another independent professional who agrees with his or her judgment.

Euthanasia in these circumstances is strongly supported by the Royal Dutch Medical Association, and by the general public in the Netherlands. The guidelines make murder in the guise of euthanasia rather far-fetched, and there is no evidence of an increase in the murder rate in the Netherlands.

It is often said, in debates about euthanasia, that doctors can be mistaken. In rare instances patients diagnosed by two competent doctors as suffering from an incurable condition have survived and enjoyed years of good health. Possibly the legalization of voluntary euthanasia would, over the years, mean the deaths of a few people who would otherwise have recovered from their immediate illness and lived for some extra years. This is not, however, the knockdown argument against euthanasia that some imagine it to be. Against a very small number of unnecessary deaths that might occur if euthanasia is legalized we must place the very large amount of pain and distress that will be suffered if euthanasia is not legalized, by patients who really are terminally ill. Longer life is not such a supreme good that it outweighs all other considerations. (If it were, there would be many more effective ways of saving life—such as a ban on smoking, or a reduction of speed limits to 25 miles per hour—than prohibiting voluntary euthanasia.) The possibility that two doctors may make a mistake means that the person who opts for euthanasia is deciding on the balance of probabilities and giving up a very slight chance of survival in order to avoid suffering that will almost certainly end in death. This may be a perfectly rational choice. Probability is the guide of life, and of death, too. Against this, some will reply that improved care for the terminally ill has eliminated pain and made voluntary euthanasia unnecessary. Elisabeth Kübler-Ross, whose *On Death and Dying* is perhaps the best-known book on care for the dying, has claimed that none of her patients request euthanasia. Given personal attention and the right medication, she says, people come to accept their death and die peacefully without pain.

Kübler-Ross may be right. It may be possible, now, to eliminate pain. In almost all cases, it may even be possible to do it in a way that leaves patients in possession of their rational faculties and free from vomiting, nausea, or other distressing side effects. Unfortunately, only a minority of dying patients now receive this kind of care. Nor is physical pain the only problem. There can also be other distressing conditions, like bones so fragile they fracture at sudden movements, uncontrollable nausea and vomiting, slow starvation due to a cancerous growth, inability to control one's bowels or bladder, difficulty in breathing, and so on.

Dr. Timothy Quill, a doctor from Rochester, New York, has described how he prescribed barbiturate sleeping pills for "Diane," a patient with a severe form of leukemia, knowing that she wanted the tablets in order to be able to end her life. Dr. Quill had known Diane for many years, and he admired her courage in dealing with previous serious illnesses. In an article in the *New England Journal of Medicine*, Dr. Quill wrote:

> It was extraordinarily important to Diane to maintain control of herself and her own dignity during the time remaining to her. When this was

> no longer possible, she clearly wanted to die. As a former director of a hospice program, I know how to use pain medicines to keep patients comfortable and lessen suffering. I explained the philosophy of comfort care, which I strongly believe in. Although Diane understood and appreciated this, she had known of people lingering in what was called relative comfort, and she wanted no part of it. When the time came, she wanted to take her life in the least painful way possible. Knowing of her desire for independence and her decision to stay in control, I thought this request made perfect sense. . . . In our discussion it became clear that preoccupation with her fear of a lingering death would interfere with Diane's getting the most out of the time she had left until she found a safe way to ensure her death.[2]

Not all dying patients who wish to die are fortunate enough to have a doctor like Timothy Quill. Betty Rollin has described, in her moving book *Last Wish,* how her mother developed ovarian cancer that spread to other parts of her body. One morning her mother said to her:

> I've had a wonderful life, but now it's over, or it should be. I'm not afraid to die, but I am afraid of this illness, what it's doing to me. . . . There's never any relief from it now. Nothing but nausea and this pain. . . . There won't be any more chemotherapy. There's no treatment anymore. So what happens to me now? I know what happens. I'll die slowly. . . . I don't want that. . . . Who does it benefit if I die slowly? If it benefits my children I'd be willing. But it's not going to do you any good. . . . There's no point in a slow death, none. I've never liked doing things with no point. I've got to end this.[3]

Betty Rollin found it very difficult to help her mother to carry out her desire: "Physician after physician turned down our pleas for help (How many pills? What kind?)." After her book about her mother's death was published, she received hundreds of letters, many from people, or close relatives of people, who had tried to die, failed, and suffered even more. Many of these people were denied help from doctors, because although suicide is legal in most jurisdictions, assisted suicide is not.

Perhaps one day it will be possible to treat all terminally ill and incurable patients in such a way that no one requests euthanasia and the subject becomes a nonissue; but this is now just a utopian ideal, and no reason at all to deny euthanasia to those who must live and die in far less comfortable conditions. It is, in any case, highly paternalistic to tell dying patients that they are now so well looked after that they need not be offered the option of euthanasia. It would be more in keeping with respect for individual freedom and autonomy to legalize euthanasia and let patients decide whether their situation is bearable.

Do these arguments for voluntary euthanasia perhaps give too much weight to individual freedom and autonomy? After all, we do not allow people free choices on matters like, for instance, the taking of heroin. This is a restriction of freedom but, in the view of many, one that can be justified on paternalistic grounds. If preventing people from becoming heroin addicts is justifiable paternalism, why isn't preventing people from having themselves killed?

The question is a reasonable one, because respect for individual freedom can be carried too far. John Stuart Mill thought that the state should never interfere with the individual except to prevent harm to others. The individual's own good, Mill thought, is not a proper reason for state intervention. But Mill may have had too high an opinion of the rationality of a human being. It may occasionally be right to prevent people from making choices that are obviously not rationally based and that we can be sure they will later regret. The prohibition of voluntary euthanasia cannot be justified on paternalistic grounds, however, for voluntary euthanasia is an act for which good reasons exist. Voluntary euthanasia occurs only when, to the best of medical knowledge, a person is suffering from an incurable and painful or extremely distressing condition. In these circumstances one cannot say that to choose to die quickly is obviously irrational. The strength of the case for voluntary euthanasia lies in this combination of respect for the preferences, or autonomy, of those who decide for euthanasia, and in the clear rational basis of the decision itself.

Euthanasia: Emerging from Hitler's Shadow

Voluntary Euthanasia and Hitler's Shadow

FOR FIFTY YEARS, Adolf Hitler has cast a long and dark shadow over discussions of euthanasia. His shadow still persists, to the extent that every time euthanasia is debated, the slippery slope appears beneath our feet, and the Holocaust lies at the bottom of it.

The most oft-quoted passage about Nazism and euthanasia comes from the American psychiatrist, Major Leo Alexander, who was commissioned to report on Nazi compulsory sterilization and the so-called "euthanasia" program. According to Alexander:

> Whatever proportions [Nazi] crimes finally assumed, it became evident to all who investigated them that they had started from small beginnings. The beginnings at first were merely a subtle shift in emphasis in the basic attitude of the physicians. It started with the acceptance of the attitude, basic in the euthanasia movement, that there is such a thing as a life not worthy to be lived. This attitude in its early stages concerned itself merely with the severely and chronically sick. Gradually the sphere of those to be included in the category was enlarged to encompass the socially unproductive, the ideologically unwanted, the racially unwanted and finally all non-Germans. But it is important to realize that the infinitely

> small wedged-in lever from which this entire trend of mind received its impetus was the attitude toward the nonrehabilitable sick.[1]

Alexander's message was simple: to prevent Nazism from recurring, we must deny that there is such a thing as "a life not worthy to be lived." The expression Alexander chose needs precise examination. It reads oddly in English, no doubt because it is a fairly literal translation of *lebensunwertes Leben*, a phrase used by the Nazis to describe those they considered not worthy of living, because they were a blot on the nobility of the Aryan *Volk*. Alexander points out that this expression was first applied to the "severely and chronically sick"—it would have been more accurate to say, to people who were intellectually disabled or incurably psychiatrically ill. The "euthanasia program" which the Nazis initiated in September 1939 focused on these people.

This Nazi "euthanasia" program was not "euthanasia" at all. It did not seek to provide a good death for human beings who were leading a miserable life. It was aimed at improving the quality of the *Volk* and eliminating the burden of caring for "social ballast" and feeding "useless mouths." That this is so is clear from the fact that war veterans were exempted from the program, and relatives were given false information about the fate of their kin. In other words, the Nazis themselves recognized that what they were doing was contrary to the interests of the people they were doing it to, and they were not willing to publicly justify their actions. It is also not at all coincidental that this program was put into action as soon as war broke out. As the historian Michael Burleigh has pointed out:

> The "euthanasia" program for adults was very much bound up with clearing the decks in order to wage war. . . . Mentally and physically disabled people were killed to save money and resources, or to create physical space for ethnic German repatriates and/or civilian and military casualties. . . . In other words, these policies reflected a nonmedical agenda, even if some psychiatrists belatedly tried to give them some *ex post facto* rationalisation in order to defend their little beleaguered professional empires. The idea of "modernizing" psychiatric provision came to them after they had cold-bloodedly murdered tens of thousands of people, while re-creating scenes of "medieval" desolation in the asylums, with skeletal patients lying on straw in vermin-infested bunkers.[2]

Alexander's view may be simply that the attitude that made the Holocaust possible was the attitude that some lives are not "worthy" of being lived because they fall below some ideal standard of racial and genetic purity. Whether this view is correct is a huge and complex historical question, but it is not my purpose to deny its validity. It may well be right. If it is, then the attitude that lies at the root of the Holocaust is, fortunately, an attitude that very few people hold today. While racism regrettably remains with us, in various ways, the defeat of Nazism was a definitive defeat for the particular variety of racism that held up one racial ideal type as embodying all the best qualities of the human species. Think, for example, of the way in which the Nazis tried to downplay the victories of Jesse Owens at the 1936 Berlin Olympics, because they were inconsistent with the idea of Aryan superiority. Can we imagine anyone today arguing that whites, or Europeans, or "Aryans," whatever they might be, are naturally superior to all other races in athletics? Can anyone argue that the European race has a monopoly on moral virtue, or reasoning skills, or artistic talent? Our world has become much more diverse and intertwined, and we draw on talent from all parts of the world in all the fields of human endeavor. There is a wide degree of acceptance that this should be so. Part of this is also an acceptance of the benefits of racial diversity, and of a mixing between people of different ethnic backgrounds, extending, very naturally, to sexual relationships and reproduction. With the hindsight given by history, we can see that the Nazi adulation of a racially pure Aryan *Volk* was the last gasp of an ideal that was already flying in the face of reality. Today, it is simply absurd.

The Idea of a Life Not Worth Living

SO THE IDEA of a life not *worthy* to be lived, in the sense of not measuring up to some standard of value, some ideal baseline that must be met for life to be sufficiently worthy of being lived, is dead and buried. But we must distinguish this from the superficially similar, but quite distinct idea, of a life not *worth* living. Was Alexander intending to suggest that to avoid starting down the road that leads to the Holocaust, we must refrain from ever judging that a person's quality of life has fallen so low that it is better for that person if he or she should die?

If this was Alexander's view, it is surely wrong. We usually value life because it is the basis for everything else that we value, whether it be happiness, appreciation of beauty, creativity, love, or the exercise of our rational faculties. But there comes a time in the lives of many people when life can no longer support these things we value, or else is so racked by pain, discomfort, nausea, or other forms of suffering that it has more negative value than positive value. An individual who is adult and of sound mind is the best judge of when his or her life has lost what is positive about it. If in the case of a terminal or incurable severe illness it is reasonable to believe that these positive qualities can never be recovered, then it can also be reasonable to regard the days, weeks, or months that are left as being of no value, or even of negative value. Many people do so regard them. They judge their own life as not worth living, and they seek to end it. If that were not the case, Jack Kevorkian would not have people begging him for assistance, Derek Humphry's *Final Exit* would not have become a best-seller, and we would not today be debating, in many different countries, the legalization of voluntary euthanasia and physician-assisted suicide.

Now even the law is starting to recognize that there is such a thing as a life not worth living. I referred earlier to the decision of the highest court in Britain, the House of Lords, in the case of Tony Bland, a young man in a persistent vegetative state. That decision, I argued, was explicitly based on the view that continued life is not always a benefit to the person living it. The law lords held, unanimously, that when this is clearly the case, it is not a criminal offense for a doctor to withdraw food and fluid, with the explicit intention of ending the patient's life. Nor is this decision an isolated one. It has subsequently been applied in another British case (*Frenchay Healthcare National Health Service Trust* v. *S* [1994] *1 WLR 601*), while a series of decisions by the British court of appeal concerning severely disabled infants has also accepted the principle that sometimes life is so "demonstrably awful" that it is not worth living.

In the Netherlands, the trial of Dr. Henk Prins has raised similar issues in the more direct and open fashion that we are now coming to expect from the Dutch. In March 1993 Dr. Prins, a Dutch gynecologist, gave a lethal injection to a four-day-old baby called Rianne. Rianne was born with a severe form of spina bifida and hydrocephalus. Clinical examination and a CT scan showed serious brain damage. Prins was part of a team of four doctors and a pastor that decided not to operate on the wound on her spine because of the severity of her handicaps. Rianne's parents agreed with this decision. Prins then discussed with the parents two possible courses to take. One was to allow the child to die slowly over a few weeks or months from untreated infections, and the other was to ensure that the child died swiftly. The parents chose the second option, and when Rianne was four days old, Prins gave her a lethal injection. He then reported what he had done to the authorities. He was prosecuted but acquitted, on the grounds that he had faced an unavoidable conflict of duties, in that he had a duty to prolong life but also a duty to alleviate unbearable suffering. The prosecution appealed, but the court of appeal in Amsterdam upheld Prins's acquittal, and the Dutch supreme court has declined to hear the case.[3]

The idea of "a life not worth living," which is implicit in these decisions, takes the viewpoint of the subject of the life. For Tony Bland, the House of Lords was saying that life was not worth living because he was permanently unable to experience anything at all. For Baby J, a brain-damaged, prematurely born infant who at five months of age was unable to breathe without a ventilator, was blind, was unable to sit up, and would probably be deaf, but was capable of feeling pain, the British court of appeal thought that life was not worth living. For Baby Rianne, the Amsterdam court of appeal held that it was not in her interests for her life to be prolonged.

It is, of course, possible to say that these developments show that we are already moving down the slippery slope toward Nazism. As an example of this position, consider a recent lecture by Klaus Dörner, given at a congress to mark the fiftieth anniversary of the Nuremberg doctors' trial, and entitled "*Wenn Ärzte nur das beste wollen* . . ." ["If doctors only want the best . . ."]. Dörner, a German psychiatrist, has published extensively on the Nazi crimes against the incurably psychiatrically ill and intellectually disabled. Most of Dörner's lecture expounds his view that the origins of these Nazi programs are to be found in the Enlightenment quest for a rational and well-ordered society, which required a solution to the "social question" of what is to be done with those who do not contribute to, or fit within, the framework of such a society. (This point of view is rather different from that of Michael Burleigh, to which I referred earlier.) Dörner then turns his attention to more recent developments in bioethics, which he finds reminiscent of the ideas that led to the Nazi crimes. He attacks the churches for bowing to medical opinion in accepting that brain death is equivalent to death. He points to the draft "European Convention on Human Rights and Biomedicine," which allows, under certain conditions, research to be carried out on human beings incapable of consent. Finally and most forcefully, he attacks the growing movement to permit the withdrawal of nutrition from people in a persistent vegetative state. Here he refers to a proposal before the German Federal Chamber of Doctors which would, by adopting Swiss guidelines on assistance in dying, allow the withdrawal of nutrition from people in a persistent vegetative state in Germany as well. "With this," he says, "it would in Germany also be possible, for the first time since 1945, to extend assistance in dying to people who are not dying." He then argues that this will flow on, perhaps first to patients suffering from Alzheimer's disease, and then to the incurably psychiatrically ill and intellectually disabled. With that, he maintains, we will have returned to Nazi ideas of assistance in dying.[4]

Dörner shows us, I believe, the choices that we face. Advancing medical technology forces us to make decisions that we did not have to make before. At the end of the Second World War there was no need to choose between those who need organ transplants to survive and those whose brains have irreversibly ceased to function but whose bodily functions are being maintained by intensive medical care. Nor were we able to scan the brains of patients in a vegetative state in order to diagnose some as having no chance of recovering consciousness. So, if like Dörner we take Hitler's shadow as the dominant factor whenever we make life-and-death judgments about others, we will have to let people in need of organ donations die because brain death is not really death. We will have to give up hope of finding better treatments for stroke victims, because we will be unable to try out these treatments on patients who have just suffered a stroke and are unable to give consent to taking part in a research program. We will have to continue to feed and care for people like Tony Bland for the rest of their natural life, which may be fifty or sixty years.

For me, those choices are not difficult, and I am not at all persuaded that the practices Dörner criticizes have any tendency to lead to Nazi-like attitudes. But there are some questions that are more difficult. Among them are questions concerning the treatment of infants with Down syndrome. Those who know and care for people with Down syndrome agree that it is a life with more limited opportunities than those available to most other people, but Down syndrome is not a condition that leads to a miserable life for the person with the syndrome. People with Down syndrome often have a happy and cheerful disposition. Hence it would be difficult to argue plausibly that ending the life of a person with Down syndrome was in the interests of that person, or that life with Down syndrome is a life not worth living.

Down syndrome happens to have been one of the first abnormalities to be detected in the fetus by means of prenatal diagnosis. In most developed countries prenatal diagnosis is now routinely offered to pregnant women over a certain age, usually thirty-five but in some countries less than that. The fact that older women are tested reflects the fact that older women are more likely to bear a child with Down syndrome. In other words, it is a policy that reflects the view that it is desirable to detect Down syndrome in the fetus. When Down syndrome is detected and abortion is available, the overwhelming majority of women, in most countries in excess of 90 percent, choose abortion. The fact that so many women carrying fetuses with Down syndrome choose not to give birth to the child surely tells us something about their attitude to life with Down syndrome, and their desire to avoid, if possible, being the mother of such a child. At the same time, the widespread acceptance of the practice of termination of pregnancy for precisely these reasons shows that we are willing to allow parents to make this choice. Indeed, it is interesting to note that the right to terminate a pregnancy after an abnormal fetus has been detected is much more widely supported than the general right to terminate a pregnancy because the woman does not wish to be pregnant. Thalidomide cases were a major impetus behind the liberalization of abortion laws in Britain, the United States, and many European countries.[5] In a recent Swedish study entitled *The Moral Roots of Prenatal Diagnosis*, Christian Munthe points out that in the Swedish debate about abortion in the early 1970s, Christian opponents of a parliamentary proposal to allow unrestricted abortion up to the eighteenth week of gestation explicitly exempted from their opposition cases of abortion where the fetus has, or probably has, a disorder or disease.[6]

Abortion tends to be seen in terms of the right of the pregnant woman to control her own body, rather than a decision about whether the prospective life of the child is worth living, and decisions about fetuses are not seen in the same light as decisions about people after birth. These issues are raised less ambiguously when life-and-death decisions must be made after the child has been born. The highly controversial Baby Doe case, which occurred in Bloomington, Indiana, in April 1982, was such a case. It involved a deliberate decision to allow an infant with Down syndrome and an intestinal blockage to die—a decision that was held to be justifiable by the parents and their family doctor and was upheld by the supreme court of Indiana, although it led the Reagan administration to try to stop the refusal of lifesaving treatment to disabled infants.[7]

It is at this point that the issue becomes most difficult. On the one hand, apart from the fact that an infant can be given up for adoption, I cannot see much difference between the widely accepted termination of a pregnancy because the fetus has Down syndrome and the much more controversial decision not to prolong the life of a baby with the same condition. And I have to agree with Dr. Henk Prins when he says: "Giving the injection was not the killing decision. The killing decision was to withhold treatment."[8] On the other hand, in a case like that of Baby Doe, what we are really talking about is ending the life of an infant, not for the sake of the infant as such, but because its life will be far more limited than the life that the parents had envisaged for their child.

A Tentative Conclusion

THE WAY IN WHICH THIS DEBATE continues shows that it has largely emerged from Hitler's shadow. So it should. I say this, not because I think we should forget about the Holocaust; on the contrary, it is important to keep reminding ourselves, and our children and grandchildren, of the nature of this terrible episode that shows, better than any other in this century, the worst of which we are capable. Yet we cannot build effective barriers against a return to the past by doing things that are plainly futile, like maintaining the lives of those who will never recover consciousness. We must strive to build a defensible ethic for these difficult questions, and so find our own way forward. The debate over euthanasia—and more generally, over medical decision-making at the end of life—is emerging from Hitler's shadow precisely to the extent that we can now move to a more open and complex discussion of what is and what is not sound in the stands we take about ethical issues regarding the end of life. Here I agree entirely with a comment made by Prins in an interview with Arlene Klotzko. Klotzko asked Prins if allowing quality of life to play a role in life-and-death decisions raised slippery-slope problems. Prins said, "Yes, but life is inevitably a slippery slope and everyone is on a slippery slope. If we do not put things out in the open—think and talk about them—the danger of the slippery slope is greater."[9]

In Place of the Old Ethic

FROM *Rethinking Life and Death*

The Structure of Ethical Revolutions

FOUR HUNDRED YEARS AGO our views about our place in the universe underwent a crisis. The ancients used a model of the solar system devised by Ptolemy, according to which the earth was the center of the universe and all the heavenly bodies revolved around it. Even the ancients knew, however, that this model did not work very well. It did not predict the positions of the planets with sufficient accuracy. So it was assumed that, as the planets moved in great circles around the earth, they also moved in smaller circles around their own orbits. This helped to patch up the model, but it didn't fix all the problems, and further adjustments were required. These adjustments were again an improvement, but still did not quite get it right. It would have been possible to add yet another modification to the basic geocentric model—but then Copernicus proposed a radically new approach. He suggested that the planets, including the earth, revolve around the sun. This remarkable new view met stiff resistance, because it required us to give up our cherished idea that we are the center of the universe. It also clashed with the Judeo-Christian view of human beings as the pinnacle of creation. If we are the reason why everything else was made, why do we have such an undistinguished address?

The resistance to the Copernican theory was not, however, simply due to human pride, hidebound conservatism, or religious prejudice. The truth is that, in predicting the movements of the planets, Copernicus was not really any more accurate than the latest patched-up version of Ptolemy's old system. For Copernicus, too, had made a mistake. He clung to the idea that the heavenly bodies move in perfect circles, when really, as Kepler was later to show, the orbits of the planets are slightly elliptical. So there were some who continued to believe in the ancient model of the universe, and looked for better ways of making it fit the facts. The Copernican theory nevertheless triumphed, not because it was more accurate than the old one, but because it was a fresh approach, full of promise.[1]

Like cosmology before Copernicus, the traditional doctrine of the sanctity of human life is today in deep trouble. Its defenders have responded, naturally enough, by trying to patch up the holes that keep appearing in it. They have redefined death so that they can remove beating hearts from warm, breathing bodies, and give them to others with better prospects, while telling themselves that they are only taking organs from a corpse. They have drawn a distinction between "ordinary" and "extraordinary" means of treatment, which allows them to persuade themselves that their decision to withdraw a respirator from a person in an irreversible coma has nothing to do with the patient's poor quality of life. They give terminally ill patients huge doses of morphine that they know will shorten their lives, but say that this is not euthanasia, because their declared intention is to relive pain. They select severely disabled infants for "nontreatment" and make sure that they die, without thinking of themselves as killing them. By denying that an individual human being comes into existence before birth, the more flexible adherents of the sanctity of life doctrine are able to put the life, health, and well-being of a woman ahead of that of a fetus. Finally, by putting a taboo on comparisons between intellectually disabled human beings and nonhuman animals, they have preserved the species boundary as the boundary of the sanctity of life ethic, despite overwhelming evidence that the differences between us and other species are differences of degree rather than of kind.

The patching could go on, but it is hard to see a long and beneficial future for an ethic as paradoxical, incoherent, and dependent on pretense as our conventional ethic of life and death has become. New medical techniques, decisions in landmark legal cases, and shifts of public opinion are constantly threatening to bring the whole edifice crashing down. Modern medical practice has become incompatible with belief in the equal value of all human life.

It is time for another Copernican revolution. It will be, once again, a revolution against a set of ideas we have inherited from the period in which the intellectual world was dominated by a religious outlook. Because it will change our tendency to see human beings as the center of the *ethical* universe, it will meet with fierce resistance from those who do not want to accept such a blow to our human pride. At first, it will have its own problems, and will need to tread carefully over new ground. For many the ideas will be too shocking to take seriously. Yet eventually the change will come. The traditional view that all human life is sacrosanct is simply not able to cope with the array of issues that we face. The new view will offer a fresh and more promising approach.

Rewriting the Commandments

WHAT WILL the new ethical outlook be like? I shall take five commandments of the old ethic that we have seen to be false, and show how they need to be rewritten for a new ethical approach to life and death. But I do not want the five new commandments to be taken as something carved in stone. I do not really approve of ethics carved in stone, anyway. There may be better ways of remedying the weaknesses of the traditional ethic. The title *Rethinking Life and Death* suggests an ongoing activity: we can rethink something more than once. The point is to start, and to do so with a clear understanding of how fundamental our rethinking must be.

First Old Commandment:
Treat all human life as of equal worth

Hardly anyone really believes that all human life is of equal worth. The rhetoric that flows so easily from the pens and mouths of popes, theologians, ethicists, and some doctors is belied every time these same people accept that we need not go all out to save a severely malformed baby; that we may allow an elderly man with advanced Alzheimer's disease to die from pneumonia, untreated by antibiotics; or that we can withdraw food and water from a patient in a persistent vegetative state. When the law sticks to the letter of this commandment, it leads to what everyone now agrees now is an absurdity, like the continuation of respirator support for Baby K, an infant born with no brain, who survived for two years, or the maintenance of Joey Fiori, a young man injured in a motorcycle accident, who was in a persistent vegetative state for almost two decades. The new approach is able to deal with these situations in the obvious way, without struggling to reconcile them with any lofty claims that all human life is of equal worth, irrespective of its potential for gaining or regaining consciousness.

First New Commandment:
Recognize that the worth of human life varies

This new commandment allows us to acknowledge frankly—as the British judges did when presented with the facts about Tony Bland's existence—that life without consciousness is of no worth at all. We can reach the same view—as British judges did in another case, *Re C,* involving an infant with a severely malformed brain—about a life that has no possibility of mental, social, or physical interaction with other human beings. Where life is not one of total or nearly total deprivation, the new ethic will judge the worth of continued life by the kind of balancing exercise recommended by Lord Justice Donaldson in a third case, *Re J,* taking into account both predictable suffering and possible compensations.

Consistent with the first new commandment, we should treat human beings in accordance with their ethically relevant characteristics. Some of these are inherent in the nature of the being. They include consciousness; the capacity for physical, social, and mental interaction with other beings; having a conscious preference for continued life; and having enjoyable experiences. Other relevant aspects depend on the relationship of the being to others: having relatives, for example, who will grieve over your death, or being so situated in a group that if you are killed, others will fear for their own lives. All of these things make a difference to the regard and respect we should have for a being.

The best argument for the new commandment is the sheer absurdity of the old one. If we were to take seriously the idea that all human life, irrespective of its capacity for consciousness, is equally worthy of our care and support, we would have to root out of medicine not only open quality-of-life judgments, but also the disguised ones. We would then be left trying to do our best to prolong indefinitely the lives of anencephalics, cortically dead infants, and patients in a persistent vegetative state. Ultimately, if we were really honest with ourselves, we would have to try to prolong the lives of those we now classify as dead because their brains have entirely ceased to function. For if human life is of equal worth, whether it has the capacity for consciousness or not, why focus on the death of the brain, rather than on the death of the body as a whole?

On the other hand, if we do accept the first new commandment, we overcome the problems that arise for a sanctity of life ethic in making decisions about anencephalics, cortically dead infants, patients in a persistent vegetative state, and those who are declared to be brain-dead by current medical criteria. In none of these cases is the really important issue how we define death. That question has had so much attention only because we are still trying to live with an ethical and legal framework formed by the old commandment. When we reject that commandment, we will instead focus on ethically relevant characteristics like the capacity for enjoyable experiences, for interacting with others, or for having a preference about continued life. Without consciousness, none of these are possible; therefore, once we are certain that consciousness has been irrevocably lost, it is not ethically relevant that there is still some hormonal brain function, for hormonal brain function without consciousness cannot benefit the patient. Nor can brain-stem function alone benefit a patient, in the absence of a cortex. So our decisions about how to treat such patients should depend not on lofty rhetoric about the equal worth of all human life, but on the views of families and partners, who deserve consideration at a time of tragic loss. If a patient in a persistent vegetative state has previously expressed wishes about what should happen to her or him in such circumstances, they should also be taken into account. (We may do this purely out of respect for the wishes of the dead, or we may do it in order to reassure others, still alive, that their wishes will not be ignored.) At the same time, in a public health-care system, we cannot ignore the limits set by the finite nature of our medical resources, or the needs of others whose lives may be saved by an organ transplant.

Second Old Commandment:
Never intentionally take innocent human life

The second commandment should be rejected because it is too absolutist to deal with all the circumstances that can arise. Taken literally, it leads to the Roman Catholic church's teaching that it is wrong to kill a fetus, *even if that would be the only way to prevent both the pregnant woman and the fetus from dying.* For those who take responsibility for the consequences of their decisions, this doctrine is absurd. It is horrifying to think that in the nineteenth and early twentieth century it was probably responsible for the preventable and agonizing death of an unknown number of women in Roman Catholic hospitals or at the hands of devout Roman Catholic doctors and midwives. This could occur if, for example, the head of the fetus became stuck during labor and could not be dislodged. Then the only way of saving the woman was to perform an operation known as a craniotomy, which involves inserting a surgical implement through the vagina and crushing the cranium, or skull, of the fetus. If this was not done, the woman and fetus would die in childbirth. Such an operation is obviously a last resort. Nevertheless, in those difficult circumstances, it seems appalling that any well-intentioned health-care professional could stand by while both woman and fetus die. For an ethic that combines an exceptionless prohibition on taking innocent human life with the doctrine that the fetus is an innocent human being, however, there could be no other course of action. If the Roman Catholic Church had said that performing a craniotomy is permissible, it would have had to give up either the absolute nature of its prohibition on taking innocent human life, or its view that the fetus is an innocent human being. Obviously, it was—and remains—willing to do neither. The teaching still stands. It is only because the development of obstetric techniques now allows the fetus to be dislodged and removed alive that the doctrine is no longer causing women to die pointlessly.

Another circumstance in which the second old commandment needs to be abandoned is—as the British law lords pointed out in deciding the Bland case—when life is of no benefit to the person living it. But the only modification to the absolute prohibition on taking human life that their lordships felt able to justify in that case—to allow a life to be taken intentionally by withholding or omitting treatment—still leaves the problem of cases in which it is better to use active means to take innocent human life. Thus, in 1992 a British court had to convict Dr. Nigel Cox of a crime because he gave Mrs. Lillian Boyes a lethal injection of potassium chloride. The fact that she had begged for death and knew that she had nothing ahead of her but a few more hours of agony, that he had tried and failed to ease her pain with large doses of morphine—all that was not enough to enable him to avoid a conviction. Needless to say, no law, no court, and no code of medical ethics would have required Dr. Cox to do everything in his power to prolong Mrs. Boyes's life. Had she suddenly become unable to breathe on her own, for instance, it would have been quite in accordance with the law and the traditional ethical view not to put her on a respirator—or if she was already on one, to take it away. The very thought of drawing out the kind of suffering that Mrs. Boyes had to endure is repugnant, and would have been regarded as wrong under the traditional ethic as well as the new one. But this only shows how much weight the traditional ethic places on the fine line between ending life by withdrawing treatment and ending it by a lethal injection. The attitude of the traditional ethic is summed up in the famous couplet:

> Thou shalt not kill; but need'st not strive
> Officiously to keep alive.

These lines are sometimes uttered in reverent tones, as if they were the wisdom of some ancient sage. One doctor, writing in the *Lancet* to defend the nontreatment of infants with spina bifida, referred to the lines as "The old dictum we were taught as medical students."[2] This is ironic, for a glance at the poem from which the couplet comes—Arthur Hugh Clough's "The Latest Decalogue"—leaves no doubt that the intention of this verse, as of each couplet in the poem, is to point out how we have failed to heed the spirit of the original ten commandments. In some of the other couplets, this is unmistakeable. For example:

> No graven images may be
> Worshipped, except the currency.

Clough would therefore have supported an extended view of responsibility. Not killing is not enough. We are also responsible for the consequences of our decision not to strive to keep alive.[3]

Second New Commandment:
Take responsibility for the consequences of your decisions

Instead of focusing on whether doctors do or do not intend to end their patients' lives, or on whether they end their patients' lives by withdrawing feeding tubes rather than giving lethal injections, the new commandment insists that doctors must ask whether a decision that they foresee will end a patient's life is the right one, all things considered.

By insisting that we are responsible for our omissions as well as for our acts—for what we deliberately don't do, as well as for what we do—we can neatly explain why the doctors were wrong to follow the Roman Catholic teaching when a craniotomy was the only way to prevent the death of both mother and fetus. But there is a price to pay for this solution to the dilemma too: unless our responsibility is limited in some way, the new ethical approach could be extremely demanding. In a world with modern means of communication and transport, in which some people live on the edge of starvation while others enjoy great affluence, there is always something that we could do, somewhere, to keep another sick or malnourished person alive. That all of us living in affluent nations, with disposable incomes far in excess of what is required to meet our needs, should be doing much more to help those in poorer countries achieve a standard of living that can meet their basic needs is a point on which most thoughtful people will agree; but the worrying aspect of this view of responsibility is that there seems to be no limit on how much we must do. If we are as responsible for what we fail to do as we are for what we do, is it wrong to buy fashionable clothes, or to dine at expensive restaurants, when the money could have saved the life of a stranger dying for want of enough to eat? Is failing to give to aid organizations really a form of killing, or as bad as killing?

The new approach need not regard failing to save as equivalent to killing. Without some form of prohibition on killing people, society itself could not survive. Society can survive if people do not save others in need—though it will be a colder, less cohesive society. Normally there is more to fear from people who would kill you than there is from people who would allow you to die. So in everyday life there are good grounds for having a stricter prohibition on killing than on allowing to die. In addition, while we can demand of everyone that he or she refrain from killing people who want to go on living, to demand too much in the way of self-sacrifice in order to provide assistance to strangers is to confront head-on some powerful and near-universal aspects of human nature. Perhaps a viable ethic must allow us to show a moderate degree of partiality for ourselves, our family, and our friends. These are the grains of truth within the misleading view that we are responsible only for what we do, and not for what we fail to do.

To pursue these questions about our responsibility to come to the aid of strangers would take us beyond the scope of this book—but two conclusions are already apparent. First, the distinction between killing and allowing to die is less clear-cut than we commonly think. Rethinking our ethic of life and death may lead us to take more seriously our failure to do enough for those whose lives we could save at no great sacrifice to our own. Second, whatever reasons there may be for preserving at least a part of the traditional distinction between killing and allowing to die—for example, maintaining that it is worse to kill strangers than to fail to give them the food they need to survive—these reasons do not apply when, like Lillian Boyes, a person wants to die, and death would come more swiftly and with less suffering if brought about by an act (for example, giving a lethal injection) than by an omission (for example, waiting until the patient develops an infection, and then not giving antibiotics).

Third Old Commandment:
Never take your own life, and always try to prevent others from taking theirs

For nearly two thousand years, Christian writers have condemned suicide as a sin. When we should die, said Thomas Aquinas, is God's decision, not ours.[4] That view became so deeply embedded in Christian nations that to attempt suicide was a crime, in some cases punished—ideologues lack a sense of irony—by death. The prohibition on suicide was one element of a general view that the state should enforce morality and act paternalistically toward its citizens.

This view of the proper role of the state was first powerfully challenged by the nineteenth-century British philosopher John Stuart Mill, who wrote in his classic *On Liberty*: "The only purpose for which power can be rightfully exercised over any member of a civilised community, against his will, is to prevent harm to others. His own good, either physical or moral, is not a sufficient warrant."[5]

Incurably ill people who ask their doctors to help them die at a time of their own choosing are not harming others. (There could be rare exceptions, for example if they have young children who need them; but people who are so ill as to want to die are generally in no position to care adequately for their children.) The state has no grounds for interfering, once it is satisfied that others are not harmed and that the decision is an enduring one, which has been freely made, on the basis of relevant information, by a competent adult person. Hence the new version of the third commandment is the direct opposite of the original version.

Third New Commandment:
Respect a person's desire to live or die

John Locke defined a "person" as a being with reason and reflection that can "consider itself as itself, the same thinking thing, in different times and places." This concept of a person is at the center of the third new commandment. Only a person can *want* to go on living, or have plans for the future, because only a person can even understand the possibility of a future existence for herself or himself. This means that to end the lives of people, against their will, is different from ending the lives of beings who are not people. Indeed, strictly speaking, in the case of those who are not people, we cannot talk of ending their lives against or in accordance with their will, because they are not capable of having a will on such a matter. To have a sense of self, and of one's continued existence over time, makes possible an entirely different kind of life. For a person, who can see her life as a whole, the end of life takes on an entirely different significance. Think about how much of what we do is oriented toward the future—our education, our developing personal relationships, our family life, our career paths, our savings, our holiday plans. Because of this, to end a person's life prematurely may render fruitless much of her past striving.

For all these reasons, killing a person against her or his will is a much more serious wrong than killing a being that is not a person. If we want to put this in the language of rights, then it is reasonable to say that only a person has a right to life.[6]

Fourth Old Commandment:
Be fruitful and multiply

This biblical injunction has been a central feature of Judeo-Christian ethics for thousands of years. The Jewish outlook regarded large families as a blessing. Augustine said that sexual intercourse without procreative intent is a sin, and to try actively to prevent procreation "turns the bridal chamber into a brothel." Luther and Calvin were equally forceful in their encouragement of procreation, with Calvin even referring to Onan's act of "spilling his seed on the ground" as "to kill before he is born the son who was hoped for." As late as 1877, the British government prosecuted Annie Besant and Charles Bradlagh for distributing a book on contraception, and around the same time in America the Comstock law prohibited the mailing or importation of contraceptives. In the twentieth century, until World War II, several European powers, among them France and Germany, continued to have national policies of increasing population in order to be able to support large armies. In some American states, old laws against the use of contraceptives survived until 1965, when the Supreme Court struck them down on the grounds that they were an invasion of privacy.[7]

Restrictions on abortion should be seen against this background view that more people are a good thing. The biblical injunction may have been apt for its time, but with world population having risen from two billion in 1930 to over six billion today, and projected to go to eleven billion by the middle of the next century, it is unethical to encourage more births. It may seem that in developed countries with low population densities, like Australia, Canada, and the United States, there is ample room for a much larger population; but all nations put their wastes down the same atmospheric and oceanic sinks, and the average Australian or North American uses several times his or her share of these sinks. If this situation continues, it will mean that people in developing countries cannot achieve a lifestyle like ours, with similar outputs of carbon dioxide and other wastes; or, if they do, then each of us will share responsibility for global warming that will speed up the melting of the polar icecaps, and so lead to a rise in sea levels. This will cause devastating floods in low-lying coastal areas, including the delta areas of Bangladesh and Egypt, where forty-six million people live. Entire island nations like Tuvalu, the Marshall Islands, and the Maldives could disappear beneath the waves.[8] Global warming may also mean devastating droughts in areas that now feed millions. Irrespective of how few people there may be per square mile, additional people living in developed countries add to the strain we are placing on the ecosystems of our planet.

The new version of the fourth commandment therefore takes a different perspective.

Fourth New Commandment:
Bring children into the world only if they are wanted

What do the original and new versions of the fourth commandment have to do with the questions discussed in this book? The two versions underpin very different views of how we should treat human life before a person comes into existence.

Consider, for example, an embryo in a laboratory. The crucial characteristic that makes it wrong to kill such an embryo, some would say, is that it has the potential to become a person, with all the characteristics that mature humans usually possess, including a degree of rationality and self-awareness that will far surpass that of a rat or a fish. But the fact that the embryo could become a person does not mean that the embryo is now capable of being harmed. The embryo does not have, and never has had, any wants or desires, so we cannot harm it by doing something contrary to its desires. Nor can we cause it to suffer. In other words the embryo is not, now, the kind of being that can be harmed, any more than the egg is before fertilization. In the absence of any meaningful sense in which the embryo can be harmed, the argument from potential seems to presuppose that it is good to promote the existence of new human beings. Otherwise, why would the fact that the embryo has a certain potential require us to realize that potential?

There are (or soon will be) as many people on this planet as it can reasonably be expected to support. If it is not wrong to kill an embryo because of the wrong it does to an existing being, then the fact that killing it will mean that one fewer person comes into existence does not make it wrong either. Those who use the potential of the embryo as an argument against abortion are rather like Calvin, who, as we saw, objected to Onan's practice of spilling his seed on the ground, because this killed "the son who was hoped for." Suppose that we really were hoping for a son. Suppose, too, that if we did not conceive a son now, then it would become impossible for us to conceive one at all. Then Calvin's objection would be sound. But if we are not hoping for another son, then the argument will pass us by.

Fifth Old Commandment:
Treat all human life as always more precious than any nonhuman life

The fifth and last of the traditional commandments that make up the sanctity of life ethic is so deeply embedded in the Western mind that even to compare human and nonhuman animals is to risk causing offence. At the time of the controversy over the Reagan administration's "Baby Doe" rules, I wrote a commentary on the issue for *Pediatrics*, the journal of the American Academy of Pediatrics. My commentary contained this sentence:

> If we compare a severely defective human infant with a nonhuman animal, a dog or a pig, for example, we will often find the nonhuman to have superior capacities, both actual and potential, for rationality, self-consciousness, communication, and anything else that can plausibly be considered morally significant.

The editor received more than fifty letters protesting against my views in this commentary, several condemning the editor for allowing it to be published. Many of the correspondents protested particularly against the comparison of the intellectual abilities of a human being and a dog or a pig. Yet the sentence that so disturbed them is not only true but *obviously* true.[9]

The lingering sense of outrage at such a comparison is a relic of the human-centred view of the universe which, as we saw, was severely battered by Copernicus and Galileo, and to which Darwin gave what ought to have been its final blow. We like to think of ourselves as the darlings of the universe. We do not like to think of ourselves as a species of animal. But the truth is that there is no unbridgeable gulf between us and other animals. Instead there is an overlap. The more intellectually sophisticated nonhuman animals have a mental and emotional life that in every significant respect equals or surpasses that of some of the most profoundly intellectually disabled human beings. This is not my subjective value-judgment. It is a statement of fact that can be tested and verified over and over again. Only human arrogance can prevent us from seeing it.

Fifth New Commandment:
Do not discriminate on the basis of species

Some people will be happy to accept the previous four new commandments but will have doubts about this one, because they associate the rejection of a bias in favor of our own species with an extreme form of species-egalitarianism that treats every living thing as of equal worth.[10] Obviously, since the new ethical outlook I have been defending rejects even the view that all *human* lives are of equal worth, I am not going to hold that *all* life is of equal worth, irrespective of its quality or characteristics. These two claims—the rejection of speciesism, and the rejection of *any* difference in the value of different living things—are quite distinct. Belief in the equal value of all life suggests that it is as wrong to uproot a cabbage as it would be to shoot dead the next person who rings your doorbell. We can reject speciesism, however, and still find many good reasons for holding that there is nothing wrong with pulling up a cabbage, while shooting the next person to ring your doorbell is utterly dreadful. For example, we can point out that cabbages lack the kind of nervous system and brain associated with consciousness, and so are not capable of experiencing anything. To uproot the cabbage therefore does not frustrate its conscious preferences for continuing to live, deprive it of enjoyable experiences, bring grief to its relatives, or cause alarm to others who fear that they too may be uprooted. To shoot the next person to ring your doorbell is likely to do all of these things.

In listing the possible reasons why it is wrong to shoot the next person to ring your doorbell, I never mentioned species. Perhaps a flying saucer has just landed in your front garden, and a friendly alien has rung the bell. If the alien is capable of having conscious preferences for continuing to live, that is a reason for not killing it. The same applies if the living alien will have enjoyable experiences, or if the alien has relatives who will grieve for its death, or other companions who will fear that they too may now be shot. So the four possible reasons I mentioned for regarding it as wrong to kill the person ringing your doorbell will apply to the alien just as they would apply to the girl from next door who wants to retrieve her ball. The rejection of speciesism implies only that the *species* of the doorbell ringer is irrelevant.

If, as I hope, this discussion has put to rest possible doubts about the new fifth commandment, it remains only to say what its implications are for the issues discussed in this book. Because membership in the species *Homo sapiens* is not ethically relevant, any characteristic or combination of characteristics that we regard as giving human beings a right to life, or as making it generally wrong to end a human life, may be possessed by some nonhuman animals. If they are, then we must grant those nonhuman animals the same right to life as we grant to human beings, or consider it as seriously wrong to end the lives of those nonhuman animals as we consider it to end the life of a human being with the same characteristic or combination of characteristics. Likewise, we cannot justifiably give more protection to the life of a human being than we give to a nonhuman animal, if the human being clearly ranks lower on any possible scale of relevant characteristics than the animal. An anencephalic baby clearly ranks lower on any possible scale of relevant characteristics than a chimpanzee. Yet as the law now stands, a surgeon could kill a chimpanzee in order to take her heart and transplant it into a human being, while to take the heart of an anencephalic baby would be murder. In terms of the revised ethical outlook, that is wrong. The right to life is not a right of members of the species *Homo sapiens;* it is—as we saw in discussing the third new commandment—a right that properly belongs to persons. Not all members of the species *Homo sapiens* are persons, and not all persons are members of the species *Homo sapiens.*

Some Answers

All we have so far is a rough sketch of how the five crumbling pillars of the old ethic might be replaced with solid new material, better able to support a structure that will guide our decisions about live and death into the next century. More thought and discussion are needed to develop these broad proposals into a working ethic. But we have to keep living in the house we are rebuilding. We cannot move out during the renovations, because decisions about life and death need to be made all the time. We cannot live without an ethic, and we cannot buy a new one ready-made. So despite the preliminary nature of our sketch of the new ethic, it is not premature to see what answers it gives to some of the issues I have discussed.

Brain Death, Anencephaly, Cortical Death, and the Persistent Vegetative State

Our examination of the sanctity of life ethic began with brain death. We saw then that the decision to regard people as dead whose brains have irreversibly ceased to function is an ethical judgment. To cease to support the bodily functions of such people is normally a justifiable ethical decision, in accordance with the first and fifth new commandments, for the most significant ethically relevant characteristic of human beings whose brains have irreversibly ceased to function is not that they are members of our species but that they have no prospect of regaining consciousness. Without consciousness, continued life cannot benefit them. There may, of course, be other issues: the need for the family to have time to adjust to a tragic loss, the preservation of organs that could save the lives of others, and occasionally a pregnancy. Exactly the same holds for patients in a persistent vegetative state, once we can be certain that there is no possibility of restoring consciousness. It holds, too—apart from the impossibility of pregnancy—for anencephalics and cortically dead infants.

This does not mean that the decision to end the life of an irreversibly unconscious patient is simple or automatic. The considerations of family feelings are both subtle and important, particularly if the patient is an infant or young person and the loss of consciousness was unexpected. But the new approach does make the decision more manageable. There are some situations in which all the considerations point in the same direction. When the parents of an anencephalic baby would like their child to be used as a source of organs to save the life of another infant, the fact that the anencephalic baby is alive should not stop us from doing the obvious thing: taking the heart from the baby who cannot benefit from continued life, and giving it to the one who can.

Abortion and the Brain-Dead Pregnant Woman

The first, fourth, and fifth new commandments have implications for the abortion controversy. What ethically relevant characteristics does the fetus have? The fact that it is a member of the species *Homo sapiens* does not answer this question. The argument from the potential of the embryo has already been examined and found wanting. In terms of the actual capacities of the fetus, there is little to suggest that it would be wrong to end its life. Probably at some point in its development in the womb the fetus does become conscious. This may happen around the tenth week of gestation, when brain activity can first be detected. Even then, brain-wave activity measurable by an EEG does not become continuous until the thirty-second week, so it may be that the fetus is only intermittently conscious until that stage, which is well past the date when it becomes viable.[11] Suppose, though, that the fetus is capable of feeling pain at the earliest possible date, ten weeks. Is the capacity to feel pain a sufficient reason to grant a being a right to life? If we think that it is, we will have to grant the same right to (at least) every normal vertebrate animal, since there is more evidence for brain activity and a capacity to feel pain even in vertebrates with relatively small brains, like frogs and fish, than there is in the fetus at ten weeks of gestation. If we balk at so radical a change in our attitudes toward nonhuman animals, we shall have to hold that the fetus may be killed for relatively trivial reasons, like those that we now consider justify us in killing rats (say, to test new food colorings) or fish (because some people prefer tuna to tofu).

An intermediate position would be that we may kill both fetuses and nonhuman animals at a similar level of awareness, if we can do so in a way that does not cause pain or distress, or if, despite the fact that some pain or distress is caused, the need to kill the fetus or nonhuman animal is sufficiently serious to outweigh the pain or distress caused. This would mean that we would have to stop the routine product safety tests now carried out on rats and other animals, because these cause the animals to become ill, and often to die in considerable distress, and the products generally do not serve any need that could not be served by an existing product. (It may not be a popular comparison to make, but rats are indisputably more aware of their surroundings, and more able to respond in purposeful and complex ways to things they like or dislike, than a fetus at ten or even at thirty-two weeks' gestation.) Fishing, too, would have to stop, except when practiced by those who have no other way of getting enough to eat. Most commercially caught fish die slowly of suffocation, as they lie gasping in the air. Recreational anglers inflict pain and distress by inducing fish to bite on a barbed metal hook. In the case of abortion, whether pain and distress is caused to the fetus would depend not only on how developed the fetus is but also on the method used. This intermediate position would allow unrestricted early abortions, and would not entirely exclude late abortions, if a method of abortion that killed the fetus painlessly were used, or if the reason for the abortion were sufficiently serious to outweigh the pain that might be caused.

It follows that there are no grounds for opposing abortion before the fetus is conscious, and only very tenuous grounds for opposing it at any stage of pregnancy. In fact, since a woman's reasons for having an abortion are invariably far more serious than the reasons most people in developed countries have for eating fish rather than tofu, and there is no reason to think that a fish suffers less when dying in a net than a fetus suffers during an abortion, the argument for not eating fish is much stronger than the argument against abortion that can be derived from the possible consciousness of the fetus after ten weeks. What has been said here of fish would also be true, in different ways, of the commercially reared and killed animals we commonly eat—quite apart from any ethically relevant characteristics that animals like cows and pigs may have in addition to their capacity to suffer. So while one may consistently be an ethical vegetarian and still accept even late abortions, those who oppose late abortions on the grounds of fetal distress will need to be ethical vegetarians if their position is to be consistent and nonspeciesist.

Resolving the issue of abortion in this way has direct implications for the dilemma with which this book began: the brain-dead woman who is pregnant. Since neither the actual characteristics of the fetus nor its potential characteristics are a reason for keeping it alive, such women can normally be allowed to become dead in every sense. The only ground for not doing so would be the strong desire of the father of the child, or of other close relatives of the pregnant woman, that the child should live. The issue then ceases to be a life-or-death decision for the fetus and becomes a question of whether the medical resources required should be used to satisfy this desire rather than others.

Infants

In the modern era of liberal abortion laws, most of those not opposed to abortion have drawn a sharp line at birth. If, as I have argued, that line does not mark a sudden change in the status of the fetus, then there appear to be only two possibilities: oppose abortion, or allow infanticide. I have already given reasons why the fetus is not the kind of being whose life must be protected in the way that the life of a person should be. Although the fetus may, after a certain point, be capable of feeling pain, there is no basis for thinking it rational or self-aware, let alone capable of seeing itself as existing in different times and places. But the same can be said of the newborn infant. Human babies are not born self-aware or capable of grasping that they exist over time. They are not persons. Hence their lives would seem to be no more worthy of protection than the life of a fetus.

Must we accept this shocking conclusion? Or does birth somehow make a difference, in some way that has so far been overlooked? Perhaps our focus on the status of the fetus and the infant has led us to neglect other aspects of the situation. Here are two ways in which birth may make a difference, not so much to the fetus or infant and its claim to life, but to others who are affected by it.

First, after birth the pregnant woman is no longer pregnant. The baby is outside her body. Thus her claim to control her own body and her own reproductive system is no longer enough to determine the life or death of the newborn baby. As I argued, this right in itself was never enough to resolve the abortion issue. Still, that does not mean that it was without any weight at all, and so the fact that at birth it no longer applies will make some difference to how we think of the newborn infant.

The second difference birth makes is that if the baby's mother does not want to keep her child, it can be cared for by someone else who does. This reason for preserving infant life is strong in a society in which there are more couples wanting to adopt a baby than there are babies needing adoption. It is no reason at all for preserving infant life if there are babies in need of adoption and no one is willing to adopt them. The coming of effective contraception and safe legal abortion have moved most developed nations sharply into the former status (though not, unfortunately, if we focus on babies with major disabilities, whom very few couples are willing to adopt). In these societies there is an important reason to protect the lives of babies, even those unwanted by their parents. In other societies that have difficulty coping with unwanted children and so have traditionally accepted infanticide, this is not a reason for preserving infant life.

So birth does make a difference to the status of the infant. But the difference is one of degree, and it remains true that the new approach, drawing on the third new commandment and the idea of a person on which that commandment is based, will not consider the newborn infant entitled to the same degree of protection as a person. There are other issues at stake as well. First, as we saw in the decisions of the British court of appeal in the cases of Baby C and Baby J, the future prospects of life may be so bleak that it is kinder to the baby, both now and in the future, to "treat it to die." That decision must depend crucially on the wishes of the parents. Their desire to keep and cherish the child can make an enormous difference to its prospects; conversely, the quality of life of a child abandoned to an institution, without loving parents, can be much less acceptable. The views of the parents, as the people most closely concerned with the infant, should also be given great weight simply because of the effects, both good and bad, that the continued life of their child will have on them and any other children they may have. In general, therefore, decisions about the future of severely disabled newborn infants should be made, not by judges who will have nothing to do with the child after their judgment is delivered, but by the parents, in consultation with their doctor.[12]

Some of the most controversial court cases have been concerned with the treatment of babies with Down syndrome. "Baby Doe" was the legal pseudonym given to a baby born in Bloomington, India, with Down syndrome and a blockage between his mouth and his stomach. Without an operation to remove the blockage, he would die. His parents refused permission for the operation. In a British case, Dr. Leonard Arthur, a respected physician, was charged with murder after putting John Pearson, a Down syndrome baby rejected by his parents, on high doses of a painkiller. People with Down syndrome have a distinctive appearance, are poorly coordinated, and are intellectually disabled to varying degrees; but they are not in pain or in need of frequent operations. Many people with Down syndrome have a cheerful temperament and can be warm and loving. In contrast to Baby C and Baby J, their lives could not be described as full of suffering, without compensating positive elements. Why then did the president of Britain's Royal College of Physicians feel able to say, at Arthur's trial, that "it is ethical that a child suffering from Down's syndrome . . . should not survive"? Why did half of a sample of American general pediatricians say that they would not oppose a parental refusal of surgery for a Down syndrome infant with blockage of the digestive system? Why did the supreme court of Indiana refuse to order that Baby Doe should be given the operation that would have saved his life?

Here is one answer to these questions. Shakespeare once described life as an uncertain voyage.[13] As parents, or would-be parents, we want our children to set out on that voyage as well equipped as possible for whatever it may bring. The expression "our children" need not refer to particular, already existing children. If we have no children but are planning to have some, we may well want to provide the children we hope to have with a good start on life's uncertain voyage. That will be better for our children—in the generic sense. But it is not *only* because we believe it will be better for our children that we may choose not to bring up a child with Down syndrome. Having children is a central part of *our* uncertain voyage as well. We will look after them and guide their lives until they are in their teens; after they become independent of us, we will still love them and share their joys and sorrows.

To have a child with Down syndrome is to have a very different experience from having a normal child. It can still be a warm and loving experience, but we must have lowered expectations of our child's abilities. We cannot expect a child with Down syndrome to play the guitar, to develop an appreciation of science fiction, to learn a foreign language, to chat with us about the latest Woody Allen movie, or to be a respectable athlete or basketball or tennis player. Even as an adult, a person with Down syndrome may not be able to live independently; and for people with Down syndrome to have children of their own is unusual and can give rise to problems. For some parents, none of this matters. They find bringing up a child with Down syndrome a rewarding experience in a thousand different ways. But for other parents, it is devastating.

Both for the sake of "our children," then, and for our own sake, we may not want a child to start on life's uncertain voyage if the prospects are clouded. When this can be known at a very early stage of the voyage, we may still have a chance to make a fresh start. This means detaching ourselves from the infant who has been born, cutting ourselves free before the ties that have already begun to bind us to our child have become irresistible. Instead of going forward and putting all our efforts into making the best of the situation, we can still say no and start again from the beginning. That is what John Pearson's mother was doing when, told that she had given birth to a Down syndrome baby, she said to her husband, "I don't want it, Duck."

It must be extraordinarily difficult to cut oneself off from one's own child and prefer it to die, so that another child with better prospects can be born. Yet many women think like this when they discover that they are pregnant with an abnormal child. Prenatal diagnosis is routine for older women, who are more at risk of having a baby with Down syndrome. It is premised on the assumption that if the test shows a fetus with Down syndrome or other abnormalities, an abortion will follow. When the pregnancy was wanted, the couple will usually then try to conceive another child.

In our culture, it is only before the baby is born that we openly accept this idea of saying no to a new life that does not have good prospects. But many other cultures say no shortly after birth as well. Among the Kung, nomadic people living in the Kalahari desert, women who give birth when they still have a child too young to walk probably do not find it easy to go to the bushes and smother the newborn infant, but doing this does not prevent them from being loving mothers to the children that they do choose to bring up. Japanese mothers are renowned for their devotion to their children, but this was compatible with the tradition of "mabiki," or "thinning" of infants. Japanese midwives who attended births did not assume that the baby was to live; instead they always asked if the baby was "to be left" or "to be returned" to wherever it was thought to have come from. Needless to say, in Japan as in all these cultures, a baby born with an obvious disability would almost always be "returned."[14]

The official western reaction to these practices is that they are shocking examples of the barbaric standards of non-Christian morality. I do not share this view. My dissent has nothing to do with cultural relativism. Some nonwestern practices—for example, female circumcision—are wrong and should, if possible, be stopped. But in the case of infanticide, it is our culture that has something to learn from others, especially now that we, like them, are in a situation where we must limit family size. I do not mean, of course, that infanticide should become a means of limiting family size. Contraception is obviously the best way to do this, since there is no point in going through an unwanted pregnancy and birth; and, for the same reason, abortion is much better than infanticide. But, for reasons we have already discussed, in regarding a newborn infant as not having the same right to life as a person, the cultures that practiced infanticide were on solid ground.

Despite the dominance of the traditional western ethic, some parents do think of their infants as replaceable. Here is a journal entry written by Peggy Stinson when Andrew, her extremely premature and seriously brain-damaged son, had been in intensive care for two months:

> I keep thinking about the other baby—the one that won't be born. The IICU [Infant Intensive Care Unit] is choosing between lives. It may already be too late for the next baby. If Andrew's life is strung out much longer, will we have the money, the emotional resources, the nerve to try again?

Another two months went by before Peggy returned to this topic in her journal. Andrew's condition had not changed.

> Thirty-fifth birthday coming up next week; haven't got forever to try for another child. . . . At this rate we'll have neither Andrew nor the next child, who because of Andrew's extended course, will have lost the chance to exist at all.
>
> Jeff [a doctor at the hospital] once said that our "next child" was theoretical, abstract—its interests couldn't be considered. Strictly speaking that may be so, but that next baby seems real enough to me. To Bob too. Decision this week to change that abstraction into a real person before it's too late.[15]

It is rare to find a couple reflecting openly on the choice between an existing baby and the "next child" whom they will conceive and raise only if the existing one dies. But there are many couples who realize that to give enough love and care to a severely disabled child would make it very difficult to bring up another child. A couple considering whether to terminate a pregnancy when the fetus has been diagnosed as having Down syndrome is in a similar situation to the Stinsons. The couple could, of course, continue the pregnancy, bring up the child with Down syndrome, and then have another child. But most couples have a sense of how many children they plan to have, and so to allow one pregnancy to lead to a child effectively precludes the existence of another child. It is implausible that the choice between one life and another does not enter the minds of many parents with disabled newborn infants. It is not Peggy Stinson's thoughts that are so unusual, but her willingness to write them down and publish them. We know that once our children's lives are properly under way, we will become committed to them; for that very reason, many couples do not want to bring up a child if they fear that both the child's life and their own experience of child rearing will be clouded by a major disability.

Shakespeare's image of life as a voyage is consistent with the idea that the seriousness of taking life increases gradually, parallel with the gradual development of the child's capacities that culminate in its life as a full person. On this view, birth marks the beginning of the next stage of development, but important changes continue to happen in the weeks and months after birth as well. These changes are not only in the capacities of the baby but also in the attachment of the parents and the acceptance of the infant into the family and the wider moral community. Many cultures have a ceremony to mark this acceptance. In ancient Greece, the infant could be exposed on the mountainside only before it had been named. (Christening may be a relic of such ceremonies.)

All of this may help to show that the ethical approach toward newborn infants proposed here is consonant with some strands of our thought about the wrongness of killing, although certainly not with all of them. Neither the first nor the second new commandment condemns the parents of Baby Doe for wanting their newborn Down-syndrome infant to die. There is no sharp ethical distinction between what they did and what most pregnant women do when they are offered an abortion because the fetus they are carrying has Down syndrome. In both cases, the decision is not primarily the concern of the state, or of the doctors—it chiefly concerns the family into whom the baby is born.

There remains, however, the problem of the lack of any clear boundary between the newborn infant, who is clearly not a person in the ethically relevant sense, and the young child, who is. In our book *Should the Baby Live?* my colleague Helga Kuhse and I suggested that a period of twenty-eight days after birth might be allowed before an infant is accepted as having the same right to life as others. This is clearly well before the infant could have a sense of its own existence over time, and would allow a couple to decide that it is better not to continue with a life that has begun very badly. The boundary is, admittedly, an arbitrary one, and this makes it problematic. We accept other arbitrary boundaries based on age, like eligibility for voting or for holding a driving licence—but a right to life is a more serious matter. Could we return to a view of infants more like that of ancient Greece, in which a public ceremony a short time after birth marked not only the parents' decision to accept the child but also society's conferral on it of the status of a person? The strongest argument for treating infants as having a right to life from the moment of birth is simply that no other line has the visibility and self-evidence required to mark the beginning of a socially recognized right to life. This is a powerful consideration; maybe in the end it is even enough to tilt the balance against a change in the law in this area. On that I remain unsure.[16]

People

The third new commandment recognizes that every person has a right to life. We have seen that the basic reason for taking this view derives from what it is to be a person, a being with awareness of her or his own existence over time, and the capacity to have wants and plans for the future.

There is also a powerful social and political reason for protecting the lives of those who are capable of fearing their own death. Universal acceptance and secure protection of the right to life of every person is the most important good that a society can bestow upon its members. Without it there is, as the seventeenth-century English philosopher Thomas Hobbes said, "continual fear and danger of violent death. And the life of man solitary, poor, nasty, brutish and short."[17] Only a being able to see herself as existing over time can fear death and can know that if people may be killed with impunity, her own life could be in jeopardy. Neither infants nor those nonhuman animals incapable of seeing themselves as existing over time can fear their own death (although they may be frightened by threatening or unfamiliar circumstances, as a fish in a net may be frightened). It is reasonable to regard more seriously crimes that cause fear in others and threaten the peaceful coexistence on which society depends. This provides another reason for recognizing that every person has a right to life, or in other words that it is a greater wrong to take the life of a person than to take the life of any other being.

A right is something one can choose to exercise or not to exercise. I have a right to a percentage of the money my publisher earns by selling this book, because I wrote it and then made an agreement with my publisher for it to be published on this basis. But I can waive this right, if I wish to do so. I could pass the royalties on to an overseas aid organization, or to the next homeless person I meet, or even tell my publisher to keep them. Similarly, the most important aspect of having a right to life is that one can choose whether or not to invoke it. We value the protection given by a right to life only when we want to go on living. No one can fear being killed at his or her own persistent, informed, and autonomous request. On the contrary, the evidence shows that many people approaching the end of their lives fear suffering much more than death. Hence the very argument that so powerfully supports recognition and protection of every person's right to life also supports the right to medical assistance in dying when this is in accordance with a person's persistent, informed, and autonomous request.

The right to medical assistance in dying has been accepted as legitimate in the Netherlands and in Oregon. Those who want to exercise it in other countries are increasingly finding ways around existing laws. But respect for a person's right to live or die also suggests that where a person is capable of expressing a view about continuing to live, life-sustaining treatment should not be withdrawn without the patient's consent. The second new commandment indicates that doctors cannot take refuge in the idea that in withdrawing treatment, they are only "letting nature take its course." On the contrary, they are responsible for the decision taken, which was to let the patient die rather than to postpone death.

The Basis of the New Approach to Life and Death

THE NEW APPROACH to life-and-death decisions is very different from the old one. But it is important to realize that the ethics of decision-making about life and death are only one part of ethics, important as they are. In particular, before leaving the sketch of the new ethic I have drawn, I want to emphasize that to deny that a being has a right to life is not to put it altogether outside the sphere of moral concern. A being that is not a person does not have the same interest in continuing to live into the future that a person usually has, but it will still have interests in not suffering and in experiencing pleasure from the satisfaction of its wants. Since neither a newborn human infant nor a fish is a person, the wrongness of killing such beings is not as great as the wrongness of killing a person. But this does not mean that we should disregard the need of an infant to be fed, and kept warm and comfortable and free of pain, for as long as it lives. Except where life is at stake, these needs should be given the weight they would be given if they were the needs of an older person. The same is true, with the necessary changes for its different needs, of the fish. Fish can surely feel pain. Their pain matters just as much, in so far as rough comparisons can be made, as similar pain experienced by a person. We do both infants and fish wrong if we cause them pain or allow them to suffer, unless to do so is the only way of preventing greater suffering.

Even when these limits to the scope of the changes I propose are understood, many will be skeptical about the need for so great a change in our ethics. There is a common view that reason and argument play no role in ethics, and therefore we have no need to defend our ethical views when they are challenged. Some people are more ready to reason about the merits of football players or chocolate cake recipes than they are about their belief in the sanctity of human life. This is a force for conservatism in ethics. It allows people to listen to a criticism of their own views and then say: "Oh, yes, well that is your opinion, but I think differently"—as if that is the end of the discussion. I hope I have shown that it is not so easy to ignore the fact that our standard view of the ethics of life and death is incoherent.

As we have just seen, the differences between the old and the new approach arise from just five key ethical commandments. In fact, the case for a drastic change to the old ethic is even simpler and more rationally compelling than that. Just as changing one or two lines of a complex computer program can completely alter the image that appears on your screen, so changing two central assumptions is enough to bring about a complete transformation of the old ethic. The first of these assumptions is that we are responsible for what we intentionally do in a way that we are not responsible for what we deliberately fail to prevent. The second is that the lives of all and only members of our species are more worthy of protection than the lives of any other beings. These are the assumptions behind the second and fifth of the commandments that we discussed earlier.

Each of these assumptions has a religious origin. The roots of the first lie in the Judeo-Christian idea of the moral law as set down in simple rules that allow for no exceptions, and the second springs from the same tradition's idea that God created man in his own image, granted him dominion over the other animals, and bestowed an immortal soul on human beings alone of all creatures. Taken independently of their religious origins, both of these crucial assumptions are on very weak ground. Can doctors who remove the feeding tubes from patients in a persistent vegetative state really believe that there is a huge gulf between this and giving the same patients an injection that will stop their hearts from beating? Doctors may be trained in such a way that it is psychologically easier for them to do the one and not the other, but both are equally certain ways of bringing about the death of the patient. As for the second assumption, what I have already said should be sufficient to show that it is not rationally defensible.

If we did nothing to the old ethic apart from abandoning these two assumptions, we would still have to construct an entirely new ethic. We could construct it differently from the ethic I have sketched out. We could, for instance, insist that, just as it is always wrong to take human life intentionally, so it is always wrong to refrain deliberately from saving human life. This would be a consistent position, but not an attractive one. It would force us to do whatever we could to keep people alive, whether they wanted to be kept alive or not and irrespective of whether they could ever recover consciousness. That would surely be a pointless and often cruel exercise of our medical powers. There would be other, even more far-reaching but on the whole much more desirable consequences for our responsibilities toward those in other countries who need food and other forms of aid that we can spare. In a similar manner, having abandoned the distinction between humans and nonhuman animals, we could refuse to adopt a distinction between persons and those who are not persons, and instead insist that every living thing or, perhaps more plausibly, every being capable of experiencing pleasure or pain has an equal right to life. So a new ethical approach can take many different forms. But without its two crucial but shaky assumptions, the old ethic cannot survive. The question is not whether it will be replaced but what the shape of its successor will be.

Ethics, Self-Interest, and Politics

The Ultimate Choice

FROM *How Are We to Live?*

Ivan Boesky's Choice

IN 1985 IVAN BOESKY WAS KNOWN as "the king of the arbitragers," arbitrage being a specialized form of investment in the shares of companies that are the target of takeover offers. He made profits of $40 million in 1981 when Du Pont bought Conoco; $80 million in 1984 when Chevron bought Gulf Oil; and in the same year, $100 million when Texaco acquired Getty Oil. There were some substantial losses too, but not enough to stop Boesky from making *Forbes* magazine's list of America's wealthiest 400 people. His personal fortune was estimated at between $150 and $200 million.[1]

Boesky had achieved both a formidable reputation and a substantial degree of respectability. His reputation came, in part, from the amount of money that he controlled. "Ivan," said one colleague, "could get any Chief Executive Officer in the country off the toilet to talk to him at seven o'clock in the morning."[2] But his reputation was also built on the belief that he had brought a new "scientific" approach to investment, based on an elaborate communications system that he claimed was like NASA's. He was featured not only in business magazines, but also in the *New York Times* Living section. He wore the best suits, on which a Winston Churchill–style gold watch chain was prominently displayed. He owned a twelve-bedroom Georgian mansion set on 190 acres in Westchester

County, outside New York City. He was a notable member of the Republican Party, and some thought he cherished political ambitions. He held positions at the American Ballet Theater and the Metropolitan Museum of Art.

Unlike other arbitragers before him, Boesky sought to publicize the nature of his work and aimed to be recognized as an expert in a specialized area that aided the proper functioning of the market. In 1985 he published a book about arbitrage entitled *Merger Mania*. The book claims that arbitrage contributes to "a fair, liquid and efficient market" and states that "undue profits are not made: there are no esoteric tricks that enable arbitragers to outwit the system. . . . Profit opportunities exist only because risk arbitrage serves an important market function." *Merger Mania* begins with a touching dedication:

> Dedication
>
> My father, my mentor, William H. Boesky (1900–1964), of beloved memory, whose courage brought him to these shores from his native Ykaterinoslav, Russia, in the year 1912. My life has been profoundly influenced by my father's spirit and strong commitment to the well-being of humanity, and by his emphasis on learning as the most important means to justice, mercy, and righteousness. His life remains an example of returning to the community the benefits he had received through the exercise of God-given talents.
>
> With this inspiration I write this book for all who wish to learn of my specialty, that they may be inspired to believe that confidence in one's self and determination can allow one to become whatever one may dream. May those who read my book gain some understanding for the opportunity which exists uniquely in this great land.[3]

In the same year that this autobiography was published, at the height of his success, Boesky entered into an arrangement for obtaining inside information from Dennis Levine. Levine, who was himself earning around $3 million annually in salary and bonuses, worked at Drexel Burnham Lambert, the phenomenally successful Wall Street firm that dominated the "junk bond" market. Since junk bonds were the favored way of raising funds for takeovers, Drexel was involved in almost every major takeover battle, and Levine was privy to information that, in the hands of someone with plenty of capital, could be used to make hundreds of millions of dollars, virtually without risk.

The ethics of this situation are not in dispute. When Boesky was buying shares on the basis of the information Levine gave him, he knew that the shares would rise in price. The shareholders who sold to him did not know that, and hence sold the shares at less than they could have obtained for them later, if they had not sold. If Drexel's client was someone who wished to take a company over, then that client would have to pay more for the company if the news of the intended takeover leaked out, since Boesky's purchases would push up the price of the shares. The added cost might mean that the bid to take over the target company would fail; or it might mean that, though the bid succeeded, after the takeover more of the company's assets would be sold off, to pay for the increased borrowings needed to buy the company at the higher price. Since Drexel, and hence Levine, had obtained the information of the intended takeover in confidence from their clients, for them to disclose it to others who could profit from it, to the disadvantage of their clients, was clearly contrary to all accepted professional ethical standards. Boesky has never suggested that he dissents from these standards or believed that his circumstances justified an exception to them. Boesky also knew that trading in inside information was illegal. Nevertheless, in 1985 he went so far as to formalize the arrangement he had with Levine, agreeing to pay him 5 percent of the profits he made from purchasing shares about which Levine had given him information.

Why did Boesky do it? Why would anyone who has $150 million, has a respected position in society, and—as is evident from the dedication to his book—values at least the appearance of an ethical life that benefits the community as a whole, risk his reputation, his wealth, and his freedom by doing something that is obviously neither legal nor ethical? Granted, Boesky stood to make very large sums of money from his arrangement with Levine. The Securities and Exchange Commission was later to describe several transactions in which Boesky had used information obtained from Levine; his profits on these deals were estimated at $50 million. Given the previous track record of the Securities and Exchange Commission, Boesky could well have thought that his illegal insider trading was likely to go undetected and unprosecuted. So it was reasonable enough for Boesky to believe that the use of inside information would bring him a lot of money with little chance of exposure. Does that mean that it was a wise thing for him to do? In these circumstances, where does wisdom lie? In choosing to enrich himself further, in a manner that he could not justify ethically, Boesky was making a choice between fundamentally different ways of living. I shall call this type of choice an "ultimate choice." When ethics and self-interest seem to be in conflict, we face an ultimate choice. How are we to choose?

Most of the choices we make in our everyday lives are restricted choices, in that they are made from within a given framework or set of values. Given that I want to keep reasonably fit, I sensibly choose to go for a walk rather than slouch on the sofa with a can of beer, watching football on television. Since you want to do something to help preserve rain forests, you join a coalition to raise public awareness of the continuing destruction of the forests. Another person wants a well-paid and interesting career, so she studies law. In each of these choices, the fundamental values are already assumed, and the choice is a matter of the best means of achieving what is valued. In ultimate choices, however, the fundamental values themselves come to the fore. We are no longer choosing within a framework that assumes that we want only to maximize our own interests, or within a framework that takes it for granted that we are going to do whatever we consider to be best, ethically speaking. Instead, we are choosing between different possible ways of living: the way of living in which self-interest is paramount, or that in which ethics is paramount, or perhaps some trade-off between the two. (I take ethics and self-interest as the two rival viewpoints because they are, in my view, the two strongest contenders. Other possibilities include, for example, living by the rules of etiquette, or living in accordance with one's own aesthetic standards, treating one's life as a work of art; but these possibilities are not the subject of this book.)

Ultimate choices take courage. In making restricted choices, our fundamental values form a foundation on which we can stand when we choose. To make an ultimate choice we must put in question the foundations of our lives. In the 1950s, French philosophers like Jean-Paul Sartre saw this kind of choice as an expression of our ultimate freedom. We are free to choose what we are to be, because we have no essential nature, that is, no given purpose outside ourselves. Unlike, say, an apple tree that has come into existence as a result of someone else's plan, we simply exist, and the rest is up to us. (Hence the name given to this group of thinkers: existentialists.) Sometimes this leads to a sense that we are standing before a moral void. We feel vertigo, and want to get out of that situation as quickly as possible. So we avoid the ultimate choice by carrying on as we were doing before. That seems the simplest and safest thing to do. But we do not really avoid making the ultimate choice in that way. We make it by default, and it may not be safe at all. Perhaps

Ivan Boesky continued to do what would make him richer because to do anything else would have involved questioning the foundations of most of his life. He acted as if his essential nature was to make money. But of course it was not: he could have chosen living ethically ahead of moneymaking.

Even if we are ready to face an ultimate choice, however, it is not easy to know how to make it. In more restricted choice situations we know how to get expert advice. There are financial consultants and educational counselors and health care advisers, all ready to tell you about what is the best for your own interests. Many people will be eager to offer you their opinions about what would be the right thing to do, too. But who is the expert here? Suppose that you have the opportunity to sell your car, which you know is about to need major repairs, to a stranger who is too innocent to have the car checked properly. He is pleased with the car's appearance, and a deal is about to be struck, when he casually asks if the car has any problems. If you say, just as casually, "No, nothing that I know of," the stranger will buy the car, paying you at least $1,000 more than you would get from anyone who knew the truth. He will never be able to prove that you were lying. You are convinced that it would be wrong to lie to him, but another $1,000 would make your life more comfortable for the next few months. In this situation you don't see any need to ask anyone for advice about what is in your best interest; nor do you need to ask what it would be right to do. So can you still ask *what to do?*

Of course you can. Some would say that if you know that it would be wrong to lie about your car, that is the end of the matter; but this is wishful thinking. If we are honest with ourselves, we will admit that, at least sometimes, where self-interest and ethics clash, we choose self-interest, and this is not just a case of being weak-willed or irrational. We are genuinely unsure what it is rational to do, because when the clash is so fundamental, reason seems to have no way of resolving it.

We all face ultimate choices, and with equal intensity, whether our opportunities are to gain, by unethical means, $50 or $50 million. The state of the world in the late twentieth century means that even if we are never tempted at all by unethical ways of making money, we have to decide to what extent we shall live for ourselves, and to what extent for others. There are people who are hungry, malnourished, lacking shelter, or lacking basic health care; and there are voluntary organizations that raise money to help these people. True, the problem is so big that one individual cannot make much impact on it; and no doubt some of the money will be swallowed up in administration, or will get stolen, or for some other reason will not reach the people who need it most. Despite these inevitable problems, the discrepancy between the wealth of the developed world and the poverty of the poorest people in developing countries is so great that if only a small fraction of what you give reaches the people who need it, that fraction will make a far greater difference to the people it reaches than the full amount you give could make to your own life. That you as an individual cannot make an impact on the entire problem seems scarcely relevant, since you can make an impact on the lives of particular families. So will you get involved with one of these organizations? Will you yourself give, not just spare change when a tin is rattled under your nose, but substantial amounts that will reduce your ability to live a luxurious lifestyle?

Some consumer products damage the ozone layer, contribute to the greenhouse effect, destroy rain forests, or pollute our rivers and lakes. Others are tested by being put, in concentrated form, into the eyes of conscious rabbits, held immobilized in rows of restraining devices like medieval stocks. There are alternatives to products that are environmentally damaging or are tested in such cruel ways. To find the alternatives can, however, be time-consuming and a nuisance. Will you take the trouble to find them?

We face ethical choices constantly in our personal relationships. We have opportunities to use people and discard them, or to remain loyal to them. We can stand up for what we believe, or make ourselves popular by going along with what the group does. Though the morality of personal relationships is difficult to generalize about because every situation is different, here too we often know what the right thing to do is, but are uncertain about what to do.

There are, no doubt, some people who go through life without considering the ethics of what they are doing. Some of these people are just indifferent to others; some are downright vicious. Yet genuine indifference to ethics of any sort is rare. Mark "Chopper" Read, one of Australia's nastiest criminals, recently published (from prison) a horrific autobiography, replete with nauseating details of beatings and forms of torture he inflicted on his enemies before killing them. Through all his relish for violence, however, the author shows evident anxiety to assure his readers that his victims were all in some way members of the criminal class who deserved what they got. He wants his readers to be clear that he has nothing but contempt for an Australian mass murderer—now one of Read's fellow-prisoners—who opened up on passersby with an automatic rifle.[4] The psychological need for ethical justification, no matter how weak that justification may be, is remarkably pervasive.

We should each ask ourselves: what place does ethics have in my daily life? In thinking about this question, ask yourself: what do I think of as a good life, in the fullest sense of that term? This is an ultimate question. To ask it is to ask: what kind of a life do I truly admire, and what kind of life do I hope to be able to look back on, when I am older and reflect on how I have lived? Will it be enough to say: "It was fun"? Will I even be able to say truthfully that it *was* fun? Whatever your position or status, you can ask what—within the limits of what is possible for you—you want to achieve with your life.

The Ring of Gyges

TWO AND A HALF THOUSAND YEARS AGO, at the dawn of Western philosophical thinking, Socrates had the reputation of being the wisest man in Greece. One day Glaucon, a well-to-do young Athenian, challenged him to answer a question about how we are to live. The challenge is a key element of Plato's *Republic*, one of the foundational works in the history of Western philosophy. It is also a classic formulation of an ultimate choice.

According to Plato, Glaucon begins by retelling the story of a shepherd who served the reigning king of Lydia. The shepherd was out with his flock one day when there was a storm and a chasm opened up in the ground. He went down into the chasm and there found a golden ring, which he put on his finger. A few days later, when sitting with some other shepherds, he happened to fiddle with the ring, and to his amazement discovered that when he turned the ring a certain way, he became invisible to his companions. Once he had made this discovery, he arranged to be one of the messengers sent by the shepherds to the king to report on the state of the flocks. Arriving at the palace, he promptly used the ring to seduce the queen, plotted with her against the king, killed him, and so obtained the crown.

Glaucon takes this story as encapsulating a common view of ethics and human nature. The implication of the story is that anyone who had such a ring would abandon all ethical standards—and, what is more, would be quite rational to do so:

> . . . No one, it is thought, would be of such adamantine nature as to abide in justice and have the strength to abstain from theft, and to keep his hands from the goods of others, when it would be in his power to steal anything he wished from the very marketplace with impunity, to enter men's houses and have intercourse with whom he would, to kill or to set free whomsoever he pleased; in short, to walk among men as a god. . . . If any man who possessed this power we have described should nevertheless refuse to do anything unjust or to rob his fellows, all who knew of his conduct would think him the most miserable and foolish of men, though they would praise him to each other's faces, their fear of suffering injustice extorting that deceit from them.[5]

Glaucon then challenges Socrates to show that this common opinion of ethics is mistaken. Convince us, he and the other participants in the discussion say to Socrates, that there are sound reasons for doing what is right—not just reasons like the fear of getting caught, but reasons that would apply even if we knew we would not be found out. Show us that a wise person who found the ring would, unlike the shepherd, continue to do what is right.

That, at any rate, is how Plato described the scene. According to Plato, Socrates convinced Glaucon and the other Athenians present that, whatever profit injustice may seem to bring, only those who act rightly are really happy. Unfortunately, few modern readers are persuaded by the long and complicated account that Socrates gives of the links between acting rightly, having a proper harmony between the elements of one's nature, and being happy. It all seems too theoretical, too contrived, and the dialogue becomes one-sided. There are obvious objections that we would like to see put to Socrates, but after the initial presentation of the challenge, Glaucon's critical faculties seem to have deserted him, and he meekly accepts every argument Socrates puts to him.

Ivan Boesky had, in the information he received from Dennis Levine, a kind of magic ring—something that could make him as close to a king as one can get in the republican, wealth-oriented United States. As it turned out, the ring had a flaw: Boesky was not invisible when he wanted to be. But was that Boesky's only mistake, the only reason why he should not have obtained and used Levine's information? The challenge that Boesky's opportunity poses to us is a modern-day version of the challenge that Glaucon put to Socrates. Can we give a better answer?

One "answer" that is really no answer at all is to ignore the challenge.

Many people do. They live and die unreflectively, without ever having asked themselves what their goals are, and why they are doing what they do. If you are totally satisfied with the life you are now living, and quite sure that it is the life you want to lead, there is no need to read further. What is to come may only unsettle you. Until you have put to yourselves the questions that Socrates faced, however, you have not *chosen* how you live.

Ethics and Self-Interest

MORE PERSONAL DOUBTS about ethics remain. To live ethically, we assume, will be hard work, uncomfortable, self-sacrificing, and generally unrewarding. We see ethics as at odds with self-interest: we assume that those who make fortunes from insider trading ignore ethics but are successfully following self-interest (as long as they don't get caught). We do the same ourselves when we take a job that pays more than another, even though it means that we are helping to manufacture or promote a product that does no good at all or actually makes people sick. On the other hand, those who pass up opportunities to rise in their career because of ethical "scruples" about the nature of the work, or who give away their wealth to good causes, are thought to be sacrificing their own interests in order to obey the dictates of ethics. Worse still, we may regard them as suckers, missing out on all the fun they could be having, while others take advantage of their futile generosity.

This current orthodoxy about self-interest and ethics paints a picture of ethics as something external to us, even as hostile to our own interests. We picture ourselves as constantly torn between the drive to advance our self-interest and the fear of being caught doing something which others will condemn, and for which we will be punished. This picture has been entrenched in many of the most influential ways of thinking in our culture. It is to be found in traditional religious ideas that promise reward or threaten punishment for good and bad behavior, but put this reward or punishment in another realm and so make it external to life in this world. It is to be found, too, in the idea that human beings are situated at the midpoint between heaven and earth, sharing in the spiritual realm of the angels but trapped also by our brutish bodily nature in this world of the beasts. The German philosopher Immanuel Kant picked up the same idea when he portrayed us as moral beings only in so far as we subordinate our natural physical desires to the commands of universal reason that we perceive through our capacity for reason. It is easy to see a link between this idea and Freud's vision of our lives as rent by the conflict between id and superego.

The same assumption of conflict between ethics and self-interest lies at the root of much modern economics. It is propagated in popular presentations of sociobiology applied to human nature. Books like Robert J. Ringer's *Looking Out for # 1*, which was on the *New York Times* best-seller list for an entire year and is still selling steadily, tell millions of readers that to put the happiness of anyone else ahead of your own is "to pervert the laws of Nature."[6] Television, in both its programs and its commercials, conveys materialist images of success that lack ethical content. As Todd Gitlin wrote in his study of American television, *Inside Prime Time:*

> . . . prime time gives us people preoccupied with personal ambition. If not utterly consumed by ambition and the fear of ending up as losers, these characters take both the ambition and the fear for granted. If not surrounded by middle-class arrays of consumer goods, they themselves are glamorous incarnations of desire. The happiness they long for is private, not public; they make few demands on society as a whole, and even when troubled they seem content with the existing institutional order. Personal ambition and consumerism are the driving forces in their lives. The sumptuous and brightly lit settings of most series amount to advertisements for a consumption-centered version of the good life, and this doesn't even take into consideration the incessant commercials, which convey the idea that human aspirations for liberty, pleasure, accomplishment, and status can be fulfilled in the realm of consumption.[7]

The message is coming over strongly, but something is wrong. Today the assertion that life is meaningless no longer comes from existentialist philosophers who treat it as a shocking discovery; it comes from bored adolescents, for whom it is a truism. Perhaps it is the central place of self-interest, and the way in which we conceive of our own interest, that is to blame here. The pursuit of self-interest, as standardly conceived, is a life without any meaning beyond our own pleasure or individual satisfaction. Such a life is often a self-defeating enterprise. The ancients knew of the "paradox of hedonism," according to which the more explicitly we pursue our desire for pleasure, the more elusive we will find its satisfaction. There is no reason to believe that human nature has changed so dramatically as to render this ancient wisdom inapplicable.

Living Ethically

FROM *How Are We to Live?*

Heroes

YAD VASHEM IS SITUATED on a hilltop outside Jerusalem. Established by the Israeli government to commemorate the victims of the Holocaust and those who came to their aid, it is a shrine, a museum, and a research center. Leading toward the museum is a long, tree-lined avenue, the Allée des Justes, or Avenue of the Righteous. Each tree commemorates a non-Jewish person who risked her or his life in order to save a Jew during the Nazi period. Only those who gave help without expectation of reward or benefit are deemed worthy of inclusion among the righteous. Before a tree is planted a special committee, headed by a judge, scrutinizes all the available evidence concerning the individual who has been suggested for commemoration. Notwithstanding this strict test, the Avenue of the Righteous is not long enough to contain all the trees that need to be planted. The trees overflow onto a nearby hillside. There are now more than 6,000 of them. There must be many more rescuers of Jews from the Nazis who have never been identified. Estimates range from 50,000 to 500,000, but we will never really know. Harold Schulweis, who started a foundation that honors and assists such people, has pointed out that there are no Simon Wiesenthals to search out those who hid, fed, and saved the hunted. Yad Vashem, with a limited budget, can play only a passive role in reviewing evidence about people nominated by survivors. Many who were helped did not, in the end, survive; others prefer not to relive painful memories and have not come forward, or in any case could not identify their rescuers.

Perhaps the most famous of those commemorated at Yad Vashem is Raoul Wallenberg. In the early years of World War II, as the Nazis extended their rule across Europe, Wallenberg was leading a comfortable life as a Swedish businessman. Since Sweden was neutral, Wallenberg traveled extensively throughout Germany and to its ally, Hungary, in order to sell his firm's line of specialty foods. But he was disturbed at what he saw and heard of the persecution of the Jews. One of his friends described him as depressed, and added, "I had the feeling he wanted to do something more worthwhile with his life." In 1944, the scarcely credible news of the systematic extermination of the Jews began to build up to such a degree that it could no longer be ignored. The American government asked the Swedish government if, as a neutral nation, it could expand its diplomatic staff in Hungary, where there were still 750,000 Jews. It was thought that a strong diplomatic staff might somehow put pressure on the nominally independent Hungarian government to resist the deportation of Hungarian Jews to Auschwitz. The Swedish government agreed. Wallenberg was asked to go. In Budapest he found that Adolf Eichmann, who had been appointed by Himmler to administer the "final solution," was determined to show his superiors just how ruthlessly efficient he could be in wiping out the Hungarian Jewish community. Wallenberg succeeded in persuading the Hungarian government to refuse Nazi pressure for further deportations of Jews, and for a brief interlude it seemed that he could return to Sweden, his mission accomplished. Then the Nazis overthrew the Hungarian government and installed in its place a puppet regime led by the Hungarian "Arrow Cross" Nazi party. The deportations began again. Wallenberg issued Swedish "protective passes" to thousands of Jews, declaring them to have connections with Sweden and to be under the protective custody of the Swedish government. At times he stood between the Nazis and their intended victims, saying that the Jews were protected by the Swedish government and the Nazis would have to shoot him first if they wanted to take them away. As the Red Army advanced on Budapest, the situation began to disintegrate. Other neutral diplomats left, but the danger remained that the Nazis and their Arrow Cross puppets would carry out a final massacre of the Jewish ghetto. Wallenberg remained in Budapest, risking falling bombs and the hatred of trigger-happy German SS and

Hungarian Arrow Cross officers. He worked to get Jews to safer hiding places, and then to let the Nazi leaders know that if a massacre took place, he would personally see to it that they were hanged as war criminals. At the end of the war, 120,000 Jews were still alive in Budapest; directly or indirectly, most of them owed their lives to Wallenberg. Tragically, when the fighting in Hungary was over, Wallenberg himself disappeared and is presumed to have been killed, not by the Germans or the Arrow Cross, but by the Soviet secret police.[1]

Oskar Schindler was, like Wallenberg, a businessman, but of very different character and background. Schindler was an ethnic German from Moravia, in Czechoslovakia. Initially enthusiastic about the Nazi cause and the incorporation of the Czech provinces into Germany, he moved into Poland after the invading Nazi armies and took over a factory in Cracow, formerly Jewish-owned, that made enamelware. As the Nazis began taking the Jews of Cracow to the death camps, Schindler protected his Jewish workers, using as a justification the claim that his factory was producing goods essential for the war effort. On the railway platforms, as Jews were being herded into the cattle-cars that would take them to the extermination camps, he would bribe or intimidate SS officials into releasing some that he said belonged to, or had skills that were needed for, his factory. He used his own money on the black market, buying food to supplement the inadequate rations his workers received. He even traveled secretly to Budapest in order to meet with members of an underground network who could get news of the Nazi genocide to the outside world. Near the end of the war, as the Russian army advanced across Poland, he moved his factory and all his workers to a new "labor camp" he constructed at Brinnlitz in Moravia. It was the only labor camp in Nazi Europe where Jews were not beaten, shot, or worked or starved to death. All of this was very risky; twice Schindler was arrested by the Gestapo but bluffed his way out of their cells. By the end of the war, at least 1,200 of Schindler's Jewish workers had survived; without Schindler they would almost certainly have died.

Schindler exemplifies the way in which people who otherwise show no signs of special distinction prove capable of heroic altruism under the appropriate circumstances. Schindler drank heavily and liked to gamble. (Once, playing cards with the brutal Nazi commandant of a forced labor camp, he wagered all his evening's winnings for the commandant's Jewish servant, saying that he needed a well-trained maid. He won, and thus saved the woman's life.) After the war Schindler had an undistinguished career, failing in a succession of business ventures, from fur breeding to running a cement works.[2]

The stories of Wallenberg and Schindler are now well known, but there are thousands of other cases of people who took risks and made sacrifices to help strangers. Those documented at Yad Vashem include: a couple in Berlin with three children who moved out of one of the two rooms of their apartment, so that a Jewish family could live in the other room; a wealthy German who lost most of his money through his efforts to help Jews; and a Dutch mother of eight who, during the winter of 1944, when food was scarce, often went hungry, and rationed her children's food too, so that their Jewish guests could survive. Samuel Oliner was a twelve-year-old boy when the Nazis decided to liquidate the ghetto of Bobowa, the Polish town in which he was living. His mother told him to run away; he escaped from the ghetto and was befriended by a Polish peasant woman who had once done some business with his father. She helped him assume a Polish identity and arranged for him to work as an agricultural laborer. Forty-five years later Oliner, then a professor at Humboldt State University in California, cowrote *The Altruistic Personality*, a study of the circumstances and characteristics of those who rescued Jews.[3]

I know from my own parents, Jews who lived in Vienna until 1938, that for each of these heroic stories there are many more that show less dramatic, but still significant, instances of altruism. In my parents' escape from Nazi Europe, the altruism of a virtual stranger proved more effective than ties of kinship. When Hitler marched into Vienna my newly wedded parents sought to emigrate; but where could they go? To obtain an entry visa, countries like the United States and Australia required that one be sponsored by a resident, who would guarantee that the new immigrants would be of good behavior and would not be a burden on the state. My father had an uncle who, several years earlier, had emigrated to the United States. He wrote seeking sponsorship. The uncle replied that he was very willing to sponsor my father, but since he had never met my mother, he was not willing to extend the sponsorship to her! In desperation my mother turned to an Australian whom she had met only once, through a mutual acquaintance, when he was a tourist in Vienna. He had not met my father at all; but he responded immediately to my mother's request, arranged the necessary papers, met my parents on the wharf when their ship arrived, and did everything he could to make them feel welcome in their new country.

Sadly, my parents' efforts to persuade their own parents to leave Vienna were not heeded with sufficient speed. My mother's father, for example, was a teacher at Vienna's leading academic high school, until the school was ordered to dismiss all Jewish teachers. Despite the loss of employment, he believed that since he was a veteran of World War I, wounded in battle and decorated for gallantry, he and his wife would be safe from any attack on their person or lives. Until 1942 my grandparents continued to live in Vienna, under increasingly difficult conditions, until they were sent to concentration camps, which only my maternal grandmother survived. Even during the grim years of the war prior to 1942, however, we know from letters that my parents received that some non-Jews visited them, to bring news and comfort. When my grandfather became nervous about possessing his ceremonial sword (because Jews had for some time been forbidden to keep weapons), a friend of my mother's hid the sword under her coat and threw it into a canal. This woman was also a schoolteacher; her refusal to join the Nazi Party cost her any chance of promotion. Non-Jewish former pupils of my grandfather continued to visit him in his flat, and one refused to accept a university chair because he would then have been compelled to support Nazi doctrines. These were not heroic, lifesaving acts, but they were also not without a certain risk. The important point, for our purposes, is that all the social pressure on these people was pushing them in the opposite direction: to have nothing to do with Jews, and certainly not to help them in any way. Yet they did what they thought right, not what was easiest to do or would bring them the most benefit.

Primo Levi was an Italian chemist who was sent to Auschwitz because he was Jewish. He survived, and wrote *If This Is a Man*, an extraordinarily telling account of his life as a slave on rations that were not sufficient to sustain life. He was saved from death by Lorenzo, a non-Jewish Italian who was working for the Germans as a civilian on an industrial project for which the labor of the prisoners was being used. I cannot do better than close this section with Levi's reflections on what Lorenzo did for him:

> In concrete terms it amounts to little: an Italian civilian worker brought me a piece of bread and the remainder of his ration every day for six months; he gave me a vest of his, full of patches; he wrote a postcard on my behalf and brought me the reply. For all this he neither asked nor accepted any reward, because he was good and simple and did not think that one did good for a reward.
>
> . . . I believe that it was really due to Lorenzo that I am alive today;

and not so much for his material aid, as for his having constantly reminded me by his presence, by his natural and plain manner of being good, that there still existed a just world outside our own, something and someone still pure and whole, not corrupt, not savage, extraneous to hatred and terror, something difficult to define, a remote possibility of good, but for which it was worth surviving.

The personages in these pages are not men. Their humanity is buried, or they themselves have buried it, under an offence received or inflicted on someone else. The evil and insane SS men, the Kapos, the politicals, the criminals, the prominents, great and small, down to the indifferent slave Häftlinge [prisoners], all the grades of the mad hierarchy created by the Germans paradoxically fraternized in a uniform internal desolation.

But Lorenzo was a man; his humanity was pure and uncontaminated, he was outside this world of negation. Thanks to Lorenzo, I managed not to forget that I myself was a man.[4]

A Green Shoot

WE MUST, of course, be thankful for the fact that today we can help strangers without dreading the knock of the Gestapo on our door. We should not imagine, however, that the era of heroism is over. Those who took part in the "velvet revolution" that overthrew communism in Czechoslovakia, and in the parallel movement for democracy in East Germany, took great personal risks and were not motivated by thoughts of personal gain. The same can be said of the thousands who turned out to surround the Russian parliament in defense of Boris Yeltsin in his resistance to the hard-liners' coup that deposed Mikhail Gorbachev. The supreme contemporary image of this kind of courage, however, comes not from Europe, but from China. It is a picture that appeared on television and in newspapers around the world: a lone Chinese student standing in front of a column of tanks rolling toward Tiananmen Square.

In liberal democracies, living an ethical life does not involve this kind of risk, but there is no shortage of opportunities for ethical commitment to worthwhile causes. My involvement in the animal liberation movement has brought me into contact with thousands of people who have made a fundamental decision on ethical grounds: they have changed their diet, given up meat, or, in some cases, abstained from *all* animal products. This is a decision that affects your life every day. Moreover, in a society in which most people continue to eat meat, becoming a vegetarian inevitably has an impact on how others think about you. Yet thousands of people have done this, not because they believe that they will be healthier or live longer on such a diet—although this may be the case—but because they became convinced that there is no ethical justification for the way in which animals are treated when they are raised for food. For example, Mrs. A. Cardoso wrote from Los Angeles:

> I received your book, *Animal Liberation*, two weeks ago . . . I thought you would like to know that overnight it changed my thinking and I instantly changed my eating habits to that of the vegetarian . . . Thank you for making me aware of our selfishness.

There have been many letters like this. Some of the writers had no particular interest in the treatment of animals before they more or less accidentally came into contact with the issue. Typical of these is Alan Skelly, a high school teacher from the Bahamas:

> As a high school teacher I was asked to become involved in the general studies taught to grade eleven. I was asked to prepare three consecutive lessons on any social topic. My wife had been given a small leaflet, 'Animal Rights,' by a child in her class. I wrote to the organization, People for the Ethical Treatment of Animals, in Washington, DC and received on hire the video 'Animal Rights.' This video has had such an impact upon my wife and I that we are now vegetarians and committed to animal liberation. They also sent me a copy of your book, *Animal Liberation* . . . Please be aware that fourteen years after the publication of your book you are responsible for the radicalization and commitment of my wife and I to animal liberation. Perhaps next month when I show PETA's video to 100 eleventh grade students I may also extend others' moral boundaries.

Some of the people who write tell me of particular difficulties they may have; how they can't get non-leather hiking boots, or see no practical alternative to killing mice that get into their house. One had a retail fur and leather shop when he became convinced that we ought not to be killing animals for their skins—he has had problems convincing his partner to change the nature of the business! Others want to know what to feed their dogs and cats, or whether I think prawns can feel pain. Some practice their new diets alone; others work together with groups trying to change the way animals are treated. A few risk their own freedom, breaking into laboratories in order to document the pain and suffering occurring there, and perhaps to release a few animals. Wherever they draw the line, they all provide significant evidence that ethical argument can change people's lives. Once they were convinced that it is wrong to rear hens in small wire cages to produce eggs more cheaply, or to put pigs in stalls too narrow for them to turn around, these people decided that they had to bring about a moral revolution in their own lives.

Animal liberation is one of many causes that rely on the readiness of people to make an ethical commitment. For two gay Americans, the cause was the outbreak of AIDS. Jim Corti, a medical nurse, and Martin Delaney, a corporate consultant, were horrified to discover that American regulations prevented their HIV-positive friends from receiving novel drugs that appeared to offer some hope for people with AIDS. They drove to Mexico, where the drugs were available, and smuggled them back into the United States. Soon they found themselves running an illegal worldwide operation, smuggling drugs and fighting government bureaucracies that sought to protect people dying from an incurable disease against drugs that were not proven safe and effective. Eventually, after taking considerable risks and doing a lot of hard work, they succeeded in changing government policies so that AIDS patients—and all those with terminal diseases—have quicker access to experimental treatments.[5]

Australia's most memorable wilderness struggle took place in 1982 and 1983, when 2,600 people sat in front of bulldozers that were being used to begin construction of a dam on the Franklin river, in southwest Tasmania. The Franklin was Tasmania's last wild river, and the dam, to be built to generate electricity, would flood dramatic gorges and rapids, obliterate Aboriginal heritage sites, destroy Huon pines that had taken 2,000 years to grow, and drown the animals that lived in the forests. The blockaders came from all over Australia, some traveling thousands of miles at their own expense from Queensland and Western Australia. They included teachers, doctors, public servants, scientists, farmers, clerks, engineers, and taxi drivers. Almost half were arrested by the police, mostly charged with trespassing. A team of twenty lawyers, all volunteers, helped with court proceedings. Nearly 450 people refused to accept bail conditions, and spent between two and twenty-six days in jail. Professor David Bellamy, the world-renowned English botanist, traveled around the world to take part in the blockade and was duly arrested. Interviewed later in the local police lockup, he said:

> It was the most uplifting thing I have ever been part of, to see such a broad cross-section of society peacefully demonstrating in quite inhospitable weather against the destruction of something they all believed in.[6]

Ethical commitment, no matter how strong, is not always rewarded; but this time it was. The blockade made the Franklin dam a national issue and contributed to the election of a federal Labor government pledged to stop it. The Franklin still runs free.

These exciting struggles exemplify one aspect of a commitment to living ethically, but to focus too much on them can be misleading. Ethics appears in our lives in much more ordinary, everyday ways. As I was writing this chapter, my mail brought me the newsletter of the Australian Conservation Foundation, Australia's leading conservation lobby group. It included an article by the foundation's fund-raising coordinator, in which he reported on a trip to thank a donor who had regularly sent donations of $1,000 or more. When he reached the address he thought something must be wrong; he was in front of a very modest suburban home. But there was no mistake: David Allsop, an employee of the state department of public works, donates 50 percent of his income to environmental causes. David had previously worked as a campaigner himself, and said he found it deeply satisfying now to be able to provide the financial support for others to campaign.[7]

There is something uplifting about ethical commitment, whether or not we share the objectives. No doubt some who read these pages will think that it is wrong to release animals from laboratories, no matter what the animals might suffer; others will think that everyone ought to abide by the decisions of the state's planning procedures on whether or not a new dam should go ahead. They may think that those who take the opposite view are not acting ethically at all. Yet they should be able to recognize the unselfish commitment of those who took part in these actions. In the abortion controversy, for example, I can acknowledge the actions of opponents of abortion as ethically motivated, even while I disagree with them about the point from which human life ought to be protected and deplore their insensitivity to the feelings of young pregnant women who are harassed when going to clinics that provide abortions.

In contrast to most of the examples given so far, I shall now consider some in which unselfish, ethical action is a much quieter, more ordinary event, but no less significant for that. Maimonides, the greatest

Jewish moral thinker of the medieval period, drew up a "Golden Ladder of Charity." The lowest level of charity, he said, is to give reluctantly; the second lowest is to give cheerfully but not in proportion to the distress of the person in need; the third level is to give cheerfully and proportionately, but only when asked; the fourth to give cheerfully, proportionately, and without being asked, but to put the gift into the poor person's hand, thus causing him to feel shame; the fifth is to give so that one does not know whom one benefits, but they know who their benefactor is; the sixth is to know whom we benefit, but to remain unknown to them; and the seventh is to give so that one does not know whom one benefits and they do not know who benefits them. Above this highly meritorious seventh level Maimonides placed only the anticipation of the need for charity, and its prevention by assisting others to earn their own livelihood so as not to need charity at all.[8] It is striking that 800 years after Maimonides graded charity in this way, many ordinary citizens take part in what he would classify as the highest possible level of charity, at least where prevention is not possible. This happens at the voluntary blood banks that are—in Britain, Australia, Canada, and many European countries—the only source of supply for the very large amount of human blood needed for medical purposes. The gift of blood is in one sense a very intimate one (the blood that is flowing in my body will later be inside the body of another); and in another sense a very remote one (I will never know who receives my blood, nor will the recipient know from whom the blood came). It is relatively easy to give blood. Every healthy person, rich or poor, can give it, without risk. Yet to the recipient, the gift can be as precious as life itself.

It is true that only a minority of the population (in Britain, about 6 percent of people eligible to donate) actually do donate.[9] It is also true that to give blood is not much of a sacrifice. It takes an hour or so, involves a slight prick, and may make you feel a little weak for the next few hours, but that is all. How many people, a skeptic might ask, would be prepared to make a *real* sacrifice so that a stranger could live?

If the willingness to undergo anesthesia and stay overnight in a hospital is enough of a real sacrifice, we now know that hundreds of thousands of people *are* prepared to do this. In recent years, bone marrow donor registries have been established in about twenty-five countries. In the United States, about 650,000 people have registered and 1,300 have donated. Figures in some other countries are comparable. For instance, in France, 63,000 have registered and 350 have donated; England has had 180,000 registrations and 700 donations to date; in Canada, 36,000 have registered, and 83 have donated; while Denmark's registrations total 10,000, with five donations. Approximately 25,000 Australians have registered on the Australian Bone Marrow Donor Registry, and at the time of writing, ten have already donated bone marrow.[10] With calm deliberation, in a situation untouched by nationalism or the hysteria of war, and with no prospect of any tangible reward, a number of ordinary citizens are prepared to go to considerable lengths to help a stranger.

We should not be surprised about this willingness to help. As the American author Alfie Kohn puts it in a cheery book called *The Brighter Side of Human Nature:*

> It is the heroic acts that turn up in the newspaper ("Man Dives into Pond to Save Drowning Child") and upstage the dozens of less memorable prosocial behaviors that each of us witnesses and performs in a given week. In my experience, cars do not spin their wheels on the ice for very long before someone stops to give a push. We disrupt our schedules to visit sick friends, stop to give directions to lost travelers, ask crying people if there is anything we can do to help. . . . All of this, it should be stressed, is particularly remarkable in light of the fact that we are socialized in an ethic of competitive individualism. Like a green shoot forcing its way up between the concrete slabs of a city sidewalk, evidence of human caring and helping defies this culture's ambivalence about—if not outright discouragement of—such activity.[11]

Countless voluntary charities depend on public donations; and most also rely on something that, for many of us, is even harder to give: our own time. American surveys indicate that nearly 90 percent of Americans give money to charitable causes, including 20 million families who give at least 5 percent of their income to charity. Eighty million Americans—nearly half the adult population—volunteer their time; they contributed a total of 15 billion hours of volunteer work in 1988.[12]

We act ethically as consumers, too. When the public learned that the use of aerosols containing CFCs damages the ozone layer, the sale of those products fell significantly, before any legal phaseout had come into effect. Consumers had gone to the trouble of reading the labels and choosing products without the harmful chemicals, even though each of them could have chosen not to be bothered. A leading advertising agency, J. Walter Thompson, surveyed American consumers in 1990 and found that 82 percent indicated that they were prepared to pay more for environmentally friendly products. Between a third and half said that they had already made some environmental choices with their spending dollars. For example, 54 percent said that they had already stopped using aerosol sprays.[13]

The Council on Economic Priorities is a United States organization that rates companies on their corporate citizenship records. The aspects rated are giving to charity, supporting the advancement of women and members of minority groups, animal testing, military contracts, community outreach, nuclear power, involvement with South Africa, environmental impact, and family benefits. The results are published annually in a paperback that has sold 800,000 copies. Presumably many of those who buy the book are interested in supporting companies that have a good record on ethical issues.

Many of the millions of customers who have helped to make The Body Shop a successful international cosmetics chain go there because they want to make sure that when they buy cosmetics, they are not supporting animal testing or causing damage to the environment. From small beginnings, the organization has grown at an average rate of 50 percent per annum, and sales are now around $150 million a year. Similarly, mutual investment funds that restrict their investments to corporations that satisfy ethical guidelines have become much more significant in the last decade, as people become concerned about the ethical impact of their investments and not only about the financial return they may gain.[14]

These examples of ethical conduct have focused on ethical acts that help strangers, or the community as a whole, or nonhuman animals, or the preservation of wilderness, because these are the easiest to identify as altruistic, and therefore as ethical. But most of our daily lives, and hence most of our ethical choices, involve people with whom we have some relationship. The family is the setting for much of our ethical decision-making; so is the workplace. When we are in long-standing relationships with people, it is less easy to see clearly whether we do what we do because it is right, or because we want, for all sorts of reasons, to preserve the relationship. We may also know that the other person will have opportunities to pay us back—to assist us, or to make life difficult for us—according to how we behave toward him or her. In such relationships, ethics and self-interest are inextricably mingled, along with love, affection, gratitude, and many other central human feelings. The ethical aspect may still be significant.

Why Do People Act Ethically?

CYNICS BELIEVE that if only we probe deeply enough, we will find that self-interest lurks somewhere beneath the surface of every ethical action. In contrast to this view, evolutionary theory, properly understood, predicts that we will be concerned for the welfare of our kin, members of our group, and those with whom we may enter into reciprocal relationships. Now we have seen that many people act ethically in circumstances that cannot be explained in any of these ways. Oskar Schindler was not furthering his own interests, or those of his kin or of his group, when he bribed and cajoled SS officers to protect Jewish prisoners from deportation to the death camps. To a successful non-Jewish German businessman, the abject and helpless Jewish prisoners of the SS would hardly have been promising subjects with whom to begin a reciprocal relationship. (Real life has unpredictable twists; as it happened, many years after the war, when Schindler was struggling to find a career for himself, some of those whose lives he had saved were able to help him; but in 1942, as far as anyone could possibly tell, the prudent thing for Schindler to do would have been to keep his mind on his business, or relax with the wine, women, and gambling that he obviously enjoyed.) Similar things can be said about other rescuers in thousands of well-documented cases. The point is sufficiently established, though, by the more humdrum example of blood donation. Since this is an institution that continues to thrive, it is easier to investigate.

Richard Titmuss, a distinguished British social researcher, published the results of a study of nearly 4,000 British blood donors in a splendid book called *The Gift Relationship*. He asked his sample of donors why they first gave blood, and why they continued to give. Overwhelmingly, people from all levels of education and income answered that they were trying to help others. Here is one example, from a young married woman who worked as a machine operator:

> You can't get blood from supermarkets and chain stores. People themselves must come forward, sick people can't get out of bed to ask you for a pint to save their life, so I came forward in hope to help somebody who needs blood.

A maintenance fitter said simply:

> No man is an island.

A bank manager wrote:

> I felt it was a small contribution that I could make to the welfare of humanity.

And a widow on a pension answered:

> Because I am fortunate in having good health myself and like to think my blood can help someone else back to health, and I felt this was a wonderful service I wanted to be part of.[15]

Aristotle suggested that we become virtuous by practicing virtue, in much the same way as we become players of the lyre, a kind of ancient harp, by playing the lyre. In some respects this seems a strange idea, but it is supported by further research on the motivation of blood donors. Professor Ernie Lightman, of the University of Toronto, surveyed 2,000 voluntary blood donors and found that their first donation was prompted by some outside event, such as an appeal from a blood bank for more donors, the fact that friends or colleagues were donating, or the convenience of a place to donate. As time passed, however, these external motivators became less significant, and "ideas such as a sense of duty and support for the work of the Red Cross, along with a general desire to help" became more important. Lightman concludes that "with repeated performance of a voluntary act over time, the sense of personal, moral obligation assumed increasing importance." Researchers at the University of Wisconsin have also studied the motivation of blood donors and found that the greater the number of donations the donors have made, the less likely they were to say that they were prompted to give by the expectations of others, and the more likely they were to say that they were motivated by a sense of moral obligation and responsibility to the community. So maybe Aristotle was right: the more we practice virtue, for whatever reason, the more likely we are to become virtuous in an inner sense as well.[16]

Altruistic action is easy to recognize as ethical, but much ethical behavior is quite compatible with regard for one's own interests. Here is one last example, this time from my own experience. As a teenager, I worked during the summer holidays in my father's office. It was a small family business, importing coffee and tea. Among the correspondence I had to read were, occasionally, letters that my father sent out to the exporters from whom he had purchased goods, reminding them that they had not yet sent him invoices for goods dispatched a considerable time ago. Sometimes it was clear, from the length of time that had elapsed, that something had slipped through the system in the "accounts payable" section of the exporter's business. If the exporters were large firms, they might never have noticed their mistake; for us, on the other hand, since we worked on gross profit margins of 3 percent, one or two "free" consignments would have made more profit than a month's normal trading. So why not, I asked my father, let the exporters look after their own problems? If they remembered to ask for their money, well and good, if they did not, better still! His reply was that that was not how decent people did business; and anyway, to send these reminders built up trust, which was vital for any business relationship and would in the long run rebound to our profit. The answer, in other words, hovered between references to an ethical ideal of how one ought to behave (what it is to be virtuous in business, one might say) and a justification in terms of long-term self-interest. Despite this ambivalence, my father was clearly acting ethically.

Ethics is everywhere in our daily lives. It lies behind many of our choices, whether personal or political or bridging the division between the two. Sometimes it comes easily and naturally to us; in other circumstances, it can be very demanding. But ethics intrudes into our conscious lives only occasionally, and often in a confused way. If we are to make properly considered ultimate choices, we must first become more aware of the ethical ramifications of the way we live. Only then is it possible to make ethics a more conscious and coherent part of everyday life.

The Good Life

FROM *How Are We to Live?*

IT IS POSSIBLE to explain, consistently with our nature as an evolved being, why it is that we are concerned for our kin, for those with whom we can establish reciprocal relationships, and to some extent for members of our own group. Now we have seen that some people help strangers, both in heroic circumstances and in more everyday ways. Does this not break the bounds of our evolved nature? How can evolutionary theory explain a sense of responsibility to make the entire world a better place? How could those who have such a sense avoid leaving fewer descendants, and thus, over time, being eliminated by the normal workings of the evolutionary process?

Here is one possible answer. Human beings lack the strength of the gorilla, the sharp teeth of the lion, the speed of the cheetah. Brain power is our specialty. The brain is a tool for reasoning, and a capacity to reason helps us to survive, to feed ourselves, and to safeguard our children. With it we have developed machines that can lift more than many gorillas, knives that are sharper than any lion's teeth, and ways of traveling that make a cheetah's pace tediously slow. But the ability to reason is a peculiar ability. Unlike strong arms, sharp teeth, or flashing legs, it can take us to conclusions that we had no desire to reach. For reason is like an escalator, leading upward and out of sight. Once we step upon it, we do not know where we will end up.[1]

A story about how Thomas Hobbes became interested in philosophy illustrates the compelling way in which reason can draw us along. Hobbes was browsing in a library when he happened to come across a copy of Euclid's *The Elements of Geometry.* The book lay open at the Forty-seventh Theorem. Hobbes read the conclusion and swore that it was impossible. So he read the proof, but this was based on a previously proved theorem. He then had to read that; and it referred him to another theorem, and so on, until eventually the chain of reasoning led back to Euclid's set of axioms, which Hobbes had to admit were so self-evident that he could not deny them. Thus reasoning alone led Hobbes to accept a conclusion that, at first sight, he had rejected. (The episode so impressed him that in his greatest work, *Leviathan,* he attempted to apply the same deductive method of reasoning to the task of defending the right of the sovereign to absolute obedience.)[2]

Reason's capacity to take us where we did not expect to go could also lead to a curious diversion from what one might expect to be the straight line of evolution. We have evolved a capacity to reason because it helps us to survive and reproduce. But if reason is an escalator, then although the first part of the journey may help us to survive and reproduce, we may go further than we needed to go for this purpose alone. We may even end up somewhere that creates tension with other aspects of our nature. In this respect, there may after all be some validity in Kant's picture of tension between our capacity to reason, and what it may lead us to see as the right thing to do, and our more basic desires. We can live with contradictions only up to a point. When the rebelling American colonists declared that all men have the right to life, liberty, and the pursuit of happiness, they may not have intended to bring about the abolition of slavery, but they laid the foundation for a process that, over almost a century, brought about that result. Slavery might have been abolished without the Declaration of Independence, or despite the Declaration, abolition might have been staved off for another decade or two; but the tension between such universal declarations of rights and the institution of slavery was not difficult to see.

Here is another example, from Gunnar Myrdal's classic study of the American race question, *An American Dilemma.* Although this book was published in 1944, long before the civil rights victories of the 1960s, Myrdal described the process of ethical reasoning that was making racist practices difficult to sustain:

> The individual . . . does not act in moral isolation. He is not left alone to manage his rationalizations as he pleases, without interference from

> outside. His valuations will, instead, be questioned and disputed . . . The feeling of need for logical consistency within the hierarchy of moral valuations—and the embarrassed and sometimes distressed feeling that the moral order is shaky—is, in its modern intensity, a rather new phenomenon.[3]

Myrdal goes on to say that the modern intensity of this need for consistency is related to increased mobility and communication, and the spread of education. Traditional and locally held ideas are challenged by the wider society and cannot withstand the appeal of the more universal values. This factor would, Myrdal predicted, lead to wider acceptance of universal values. He was thinking of the universal application of moral principles to all within the human species; but if he were writing today, he might well consider, as a further instance of the tendency he described, the view that the interests of nonhuman animals should also receive equal consideration.[4]

Curiously, when Karl Marx wrote about the history of class revolutions, he pointed to much the same tendency:

> Each new class which displaces the one previously dominant is forced, simply to be able to carry out its aim, to represent its interest as the common interest of all members of society, that is, ideally expressed. It has to give its ideas the form of universality and represent them as the only rational, universally valid ones. . . . Every new class, therefore, achieves dominance only on a broader basis than that of the previous class ruling.[5]

Marx thought that reason was here merely providing a cloak for the class interests of those making the revolutions. Given his materialist view of history, he could hardly say anything else. Yet he also pointed out that because capitalism needed to concentrate workers in industrial centers and give them at least a minimum level of education, it contributed to raising the workers' awareness of their own situation. The same events can be seen in a different way: as the working out of the inherently universalizing nature of reasoning in societies that increasingly consist of educated and self-aware people, gradually freeing themselves from the constraints of parochial and religious beliefs. Since the general level of education and ease of communication are still increasing throughout the world, we have some grounds for hoping that this process will continue, eventually bringing with it a fundamental change in our ethical attitudes.

Our ability to reason, then, can be a factor in leading us away from both arbitrary subjectivism and an uncritical acceptance of the values of our community. The idea that everything is subjective or, more specifically, relative to our community, seems to go into and out of vogue with each generation. Like its predecessors, the postmodernist mode of relativism fails to explain how it is that we can conduct coherent discussions about the values our community should hold, or maintain that our own values are superior to those of communities that accept slavery, the genital mutilation of women, or death sentences for writers who are deemed disrespectful of the prevailing religion. In contrast, the view I have defended accounts for the possibility of this kind of discussion on the basis of two simple premises. The first is the existence of our ability to reason. The second is that, in reasoning about practical matters, we are able to distance ourselves from our own point of view and take on, instead, a wider perspective, ultimately even the point of view of the universe.

Reason makes it possible for us to see ourselves in this way because, by thinking about my place in the world, I am able to see that I am just one being among others, with interests and desires like others. I have a personal perspective on the world, from which my interests are at the front and center of the stage, the interests of my family and friends are close behind, and the interests of strangers are pushed to the back and sides. But reason enables me to see that others have similarly subjective perspectives, and that from "the point of view of the universe" my perspective is no more privileged than theirs. Thus my ability to reason shows me the possibility of detaching myself from my own perspective and shows me what the universe might look like if I had no personal perspective.

Taking the point of view of the universe as the basis of an ethical point of view does not mean that one must act impartially at all times. Some forms of partiality are themselves capable of impartial justification. For example, it is probably best for children generally if parents are regarded as having a much stricter duty to take care of their own children than they have to take care of the children of strangers. In this way society takes advantage of the natural tie of love between parents and children, which in normal circumstances is always to be preferred to the benevolence of a department of child welfare, no matter how well-intentioned the bureaucrats and social workers who make up the department may be. Love for one's children is a force that can be used for the good of all, but it does sometimes lead people to choose what is, from an impartial viewpoint, a lesser good. If the school your child attends is on fire, and you must choose between breaking open the door of the room in which she alone is trapped, and the door of another room in which twenty children are trapped—you have no time to get both doors open—most parents would probably rescue their own child. The parents of the other children might blame them for doing so, but if they were fair, they would probably recognize that in similar circumstances they would have done the same. If we weigh up the rescue of one's own child directly from an impartial standpoint, we will judge it to be wrong; but if we consider, first, the desirability of parental love for children, and then second that this act was motivated by that love, we will be more ready to accept it.[6]

Consistently with the idea of taking the point of view of the universe, the major ethical traditions all accept, in some form or other, a version of the golden rule that encourages equal consideration of interests. "Love your neighbor as yourself," said Jesus. "What is hateful to you do not do to your neighbor," said Rabbi Hillel. Confucius summed up his teaching in very similar terms: "What you do not want done to yourself, do not do to others." The *Mahabharata*, the great Indian epic, says: "Let no man do to another that which would be repugnant to himself."[7] The parallels are striking. Although Jesus and Hillel drew on a common Jewish tradition, Confucius and the *Mahabharata* appear to have reached the same position independently of each other and of the Judeo-Christian tradition. In each case, moreover, the words are offered as a kind of summary of all the moral law. Although the way in which Jesus and Hillel put the rule might be taken to limit it to members of one's own group, the parable of the good Samaritan firmly dispels this reading of whom Jesus thought one's neighbor to be.[8] Nor should Hillel, Confucius, or the *Mahabharata* be interpreted as promoting, at least in these passages, anything less than a universal ethic.

The possibility of taking the point of view of the universe overcomes the problem of finding meaning in our lives, despite the ephemeral nature of human existence when measured against all the eons of eternity. Suppose that we become involved in a project to help a small community in a developing country to become free of debt and self-sufficient in food. The project is an outstanding success, and the villagers are healthier, happier, better educated, and more economically secure, and they have fewer children. Now someone might say: "What good have you done? In a thousand years these people will all be dead, and their children and grandchildren as well, and nothing that you have done will make any difference." That may be true, or it may be false. The changes we make today could snowball and, over a long period of time, lead to much more far-reaching changes. Or they could come to nothing. We simply cannot tell. We should not, however, think of our efforts as wasted unless they endure forever, or even for a very long time. If we regard time as a fourth dimension, then we can think of the universe, throughout all the times at which it contains sentient life, as a four-dimensional entity. We can then make that four-dimensional world a better place by causing there to be less pointless suffering in one particular place, at one particular time, than there would otherwise have been. As long as we do not thereby increase suffering at some other place or time, or cause any other comparable loss of value, we will have had a positive effect on the universe.

I have been arguing against the view that value depends entirely on my own subjective desires. Yet I am not defending the objectivity of ethics in the traditional sense. Ethical truths are not written into the fabric of the universe: to that extent the subjectivist is correct. If there were no beings with desires or preferences of any kind, nothing would be of value and ethics would lack all content. On the other hand, once there are beings with desires, there are values that are not only the subjective values of each individual being. The possibility of being led, by reasoning, to the point of view of the universe provides as much "objectivity" as there can be. When my ability to reason shows me that the suffering of another being is very similar to my own suffering and (in an appropriate case) matters just as much to that other being as my own suffering matters to me, then my reason is showing me something that is undeniably *true*. I can still choose to ignore it, but then I can no longer deny that my perspective is a narrower and more limited one than it could be. This may not be enough to yield an objectively true ethical position. (One can always ask: what is so good about having a broader and more all-encompassing perspective?) But it is as close to an objective basis for ethics as there is to find.

The perspective on ourselves that we get when we take the point of view of the universe also yields as much objectivity as we need if we are to find a cause that is worthwhile in a way that is independent of our own desires. The most obvious such cause is the reduction of pain and suffering, wherever it is to be found. This may not be the only rationally grounded value, but it is the most immediate, pressing, and universally agreed upon one. We know from our own experience that when pain and suffering are acute, all other values recede into the background. If we take the point of view of the universe, we can recognize the urgency of doing something about the pain and suffering of others, before we even consider promoting (for their own sake rather than as a means to reducing pain and suffering) other possible values like beauty, knowledge, autonomy, or happiness.

Does the possibility of taking the point of view of the universe mean that the person who acts only from a narrow perspective—for the sake of self, family, friends, or nation, in ways that cannot be defended even indirectly from an impartial perspective—is necessarily acting irrationally? Not, I think, in the full sense of the term. In this respect practical reasoning—that is, reasoning about what to do—is different from theoretical reasoning. If Hobbes had accepted Euclid's axioms and been unable to find any flaw with the chain of reasoning that led from them to Euclid's Forty-seventh Theorem, but had nevertheless continued to hold that the theorem was "impossible," we could rightly have said that he had failed to grasp the nature of Euclid's reasoning process. He would simply have been in error—and if, for example, he had applied this belief to some practical problem of measurement or construction, he would have gotten the wrong answer, and this would have handicapped him in reaching whatever goal he intended the measurement or construction to achieve. If, on the other hand, I act in a way that shows less concern for the suffering of strangers than for the suffering of my family or friends, I do not show that I am incapable of grasping the point of view of the universe; I show only that this perspective does not motivate me as strongly as my more personal perspective. If to be irrational is to make a mistake, there is no mistake here; my pursuit of the more limited perspective will not lead me to a wrong answer that will prevent me from reaching my own limited objectives. We have evolved as beings with particularly strong desires to protect and further the interests of members of our family. To disregard this side of our nature altogether is scarcely possible. The most that the escalator of reason can require is that we keep it in check and remain aware of the existence of the wider perspective. So it is only in an extended sense of the term that those who take the narrower perspective might be said to be acting less rationally than those who are able to act from the point of view of the universe.

It would be nice to be able to reach a stronger conclusion than this about the basis of ethics. As things stand, the clash between self-interest and generalized benevolence has been softened, but it has not been dissolved.

Toward an Ethical Life

In a society in which the narrow pursuit of material self-interest is the norm, the shift to an ethical stance is more radical than many people realize. In comparison with the needs of people starving in Somalia, the desire to sample the wines of the leading French vineyards pales into insignificance. Judged against the suffering of immobilized rabbits having shampoos dripped into their eyes, a better shampoo becomes an unworthy goal. The preservation of old-growth forests should override our desire to use disposable paper towels. An ethical approach to life does not forbid having fun or enjoying food and wine, but it changes our sense of priorities. The effort and expense put into buying fashionable clothes, the endless search for more and more refined gastronomic pleasures, the astonishing additional expense that marks out the prestige car market from the market in cars for people who just want a reliable means of getting from A to B—all these become disproportionate to people who can shift perspective long enough to take themselves, at least for a time, out of the spotlight. If a higher ethical consciousness spreads, it will utterly change the society in which we live.

We cannot expect that this higher ethical consciousness will become universal. There will always be people who don't care for anyone or anything, not even for themselves. There will be others, more numerous and more calculating, who earn a living by taking advantage of others, especially the poor and the powerless. We cannot afford to wait for some coming glorious day when everyone will live in loving peace and harmony with everyone else. Human nature is not like that at present, and there is no sign of its changing sufficiently in the foreseeable future. Since reasoning alone proved incapable of fully resolving the clash between self-interest and ethics, it is unlikely that rational argument will persuade every rational person to act ethically. Even if reason had been able to take us further, we would still have had to face the reality of a world in which many people are very far from acting on the basis of reasoning of any kind, even crudely self-interested reasoning. So for a long time to come, the world is going to remain a tough place in which to live.

Nevertheless, we are part of this world and there is a desperate need to do something *now* about the conditions in which people live and die, and to avoid both social and ecological disaster. There is no time to focus our thoughts on the possibility of a distant utopian future. Too many humans and nonhuman animals are suffering now, the forests are going too quickly, population growth is still out of control, and if we do not bring greenhouse gas emissions down rapidly, the lives and homes of 46 million people are at risk in the Nile and Bengal delta regions alone. Nor can we wait for governments to bring about the change that is needed. It is not in the interests of politicians to challenge the fundamental assumptions of the society they have been elected to lead. If 10 percent of the population were to take a consciously ethical outlook on life and act accordingly, the resulting change would be more significant than any change of government. The division between an ethical and a selfish approach to life is far more fundamental than the difference between the policies of the political right and the political left.

We have to take the first step. We must reinstate the idea of living an ethical life as a realistic and viable alternative to the present dominance of materialist self-interest. If a critical mass of people with new priorities were to emerge, and if these people were seen to do well, in every sense of the term—if their cooperation with each other brings reciprocal benefits, if they find joy and fulfillment in their lives—then the ethical attitude will spread, and the conflict between ethics and self-interest will have been shown to be overcome, not by abstract reasoning alone, but by adopting the ethical life as a practical way of living and showing that it works, psychologically, socially, and ecologically.

Anyone can become part of the critical mass that offers us a chance of improving the world before it is too late. You can rethink your goals and question what you are doing with your life. If your present way of living does not stand up against an impartial standard of value, then you can change it. That might mean quitting your job, selling your house, and going to work for a voluntary organization in India. More often, the commitment to a more ethical way of living will be the first step of a gradual but far-reaching evolution in your lifestyle and in your thinking about your place in the world. You will take up new causes and find your goals shifting. If you get involved in your work, money and status will become less important. From your new perspective, the world will look different. One thing is certain: you will find plenty of worthwhile things to do. You will not be bored or lack fulfillment in your life. Most important of all, you will know that you have not lived and died for nothing, because you will have become part of the great tradition of those who have responded to the amount of pain and suffering in the universe by trying to make the world a better place.

Darwin for the Left

FROM *Prospect*

THE LEFT NEEDS a new paradigm. The collapse of communism and the abandonment by democratic socialist parties of the traditional socialist aim of public ownership have deprived the left of the goals it cherished over the two centuries in which it grew to a position of great political power and intellectual influence. My focus here is not so much on the left as a politically organized force as on the left as a broad body of thought, a spectrum of ideas about achieving a better society. The left, in this sense, is urgently in need of new ideas. I want to suggest that one source of such ideas is an approach to human behavior based firmly on a modern understanding of human nature. It is time for the left to take seriously the fact that we have evolved from other animals; we bear the evidence of this inheritance, not only in our anatomy and our DNA, but in what we want and how we are likely to try to get it. In other words, it is time to develop a Darwinian left.

Can the left adopt Darwin and still remain left? That depends on what it considers essential. Let me answer this in a personal way. During the past year, I have completed both a television documentary and a book about Henry Spira. This name will mean nothing to most people, but Spira is the most remarkable person I have ever worked with. When he was twelve, his family lived in Panama. His father ran a small store, which was not doing well; to save money the family accepted an offer from a rich friend to stay in his house. The house was a mansion that took up an entire city block. One day, two men who worked for the owner asked Henry if he wanted to come with them when they collected rents. He went and saw how the luxurious existence of his father's benefactor was financed; they went into the slums, where poor people were menaced by the armed rent collectors. At the time, Henry had no concept of "the left," but from that day on, he was part of it. Later Spira moved to the United States, became a Trotskyist, worked as a seaman, was blacklisted during the McCarthy era, went to the South to support black people, left the Trotskyists because they had lost touch with reality, and taught ghetto kids in New York. As if that wasn't enough, in 1973 he read my essay "Animal Liberation" and decided that here was another group of exploited beings that needed help. He has subsequently become the single most effective activist of the animal rights movement in the United States.

Spira has a knack for putting things plainly. When I asked him why he has spent his life working for these causes, he said simply that he is on the side of the weak, not the powerful, of the oppressed, not the oppressor, of the ridden, not the rider. He spoke of the huge quantity of pain and suffering that exists in our universe, and of his desire to do something to reduce it. This, I think, is what the left is about. If we shrug our shoulders at the avoidable suffering of the weak and the poor, of those who are getting exploited and ripped off, we are not of the left. The left wants to change this situation. There are many different ideas of equality which are compatible with this broad picture of the left. But in a world in which the 400 richest people have a combined net worth greater than the bottom 45 per cent of the world's population, it is not hard to find some common ground on working toward a more equal distribution of resources.

So much for the left. What about the politics of Darwinism? One way of answering the question is to invoke the fact-value distinction. Since to be "of the left" is to hold certain values, and Darwin's theory is a scientific one, the impossibility of deducing values from facts means that evolution has nothing to do with left or right. So there can be a Darwinian left as easily as there can be a Darwinian right.

It is, of course, the right which has drawn most on Darwinian thinking. Andrew Carnegie, for example, appealed to evolution to suggest that economic competition will lead to the "survival of the fittest," and so will make most people better off. Darwinian thinking is also invoked in the claim that social policies may be helping the "less fit" to survive, and thus have deleterious genetic consequences. This claim is highly speculative.

Its factual basis is strongest in regard to the provision of lifesaving medical treatment to people with genetically linked diseases; without treatment, these people would die before they could reproduce. There are, no doubt, many more people with early-onset diabetes being born because of the discovery of insulin. But no one would seriously propose withholding insulin from children with diabetes in order to avoid the genetic consequences of providing insulin.

But there is a more general aspect of Darwinian thinking that does need to be taken seriously. That is the claim that an understanding of human nature in the light of evolutionary theory can help us to assess the price we will have to pay for achieving our social and political goals. This does not imply that any social policy is wrong because it is contrary to Darwinian ideas. Rather, it leaves the ethical decision up to us and merely provides information relevant to that decision.

THE CORE OF THE LEFT WORLDVIEW is a set of values; but there is also a penumbra of factual beliefs that have typically been associated with the left. We need to ask whether these factual beliefs are at odds with Darwinian thinking; if they are, what would the left be like without them?

The intellectual left, and Marxists in particular, have generally been enthusiastic about Darwin's account of the origin of species, as long as its implications for human beings are confined to anatomy and physiology. Marx's materialist theory of history implies that there is no fixed human nature. It changes with each new mode of production. It has already changed in the past—between primitive communism and feudalism, and between feudalism and capitalism—and it could change again in the future.

Belief in the malleability of human nature has been important for the left because it has provided grounds for hoping that a different kind of society is possible. The real reason why the left rejected Darwinism is that it dashed the left's great dream: the perfectibility of man. Even before Plato's *Republic,* the idea of building a perfect society had been present in the western consciousness. For as long as the left has existed, it has sought a society in which all human beings live harmoniously and cooperatively with each other in peace and freedom. For Darwin, on the other hand, the struggle for existence, or at least for the existence of one's offspring, is unending.

In the twentieth century, the dream of the perfectibility of humankind turned into the nightmares of Stalinist Russia, China during the Cultural Revolution, and Cambodia under Pol Pot. From these nightmares the left awoke in turmoil. There have been attempts to create a new and better society with less terrible results—Castro's Cuba, the Israeli kibbutzim—but none are unqualified successes. The dream of perfectibility should be put behind us, and with that one barrier to a Darwinian left has been removed.

But what about the malleability of human nature? What do we mean by malleability, and how essential is it to the left? Let us divide human behavior into three categories: that which shows great variation across cultures, some variation across cultures, and little or no variation across cultures.

In the first category, showing great variation, I would include the way we produce our food—by gathering and hunting, by grazing domestic animals, or by growing crops. To these differences would correspond differences in lifestyles—nomadic or settled—as well as differences in the kinds of food we eat. This first category would also include economic structures, religious practices, and forms of government—but not, significantly, the existence of some form of government, which seems to be nearly universal.

In the second category, showing some variation, I would include sexuality. Victorian anthropologists were very much impressed by the differences between attitudes to sexuality in their own society and those in the societies they studied; as a result, we tend to exaggerate the extent to which sexual morality is relative to culture. There are, of course, important differences between societies that allow men to have one wife and those that allow men to have more than one wife; but almost every society has a system of marriage that implies restrictions on sexual intercourse outside marriage. Moreover, while men may be allowed one wife or more, according to the culture, systems of marriage in which women are allowed more than one husband are rare. Whatever the rules of marriage may be, and no matter how severe the sanctions, infidelity and sexual jealousy seem to be universal elements of human sexual behavior.

In this second category I would also include ethnic identification and its opposite, xenophobia and racism. I live in a multicultural society with a relatively low level of racism; but I know that racist feelings do exist among Australians, and they can be stirred up by demagogues. The tragedy of Bosnia has shown how ethnic hatred can be revived among people who have lived peacefully with each other for decades. Racism can be learned and unlearned, but racist demagogues hold their torches over highly flammable material.

In the third category, showing little variation across cultures, I would include the fact that we are social beings and that we are concerned for the interests of our kin. Our readiness to form cooperative relationships and to recognize reciprocal obligations is another universal. More controversially, I would claim that the existence of a hierarchy or system of rank is an almost universal human tendency. There are very few human societies without differences in social status; when attempts are made to abolish such differences, they tend to reemerge rapidly. Finally, gender roles also show relatively little variation. Women almost always play the main role in caring for young children, while men are much more likely than women to be involved in physical conflict, both within the social group and in warfare between groups. Men also tend to have a disproportionate role in the political leadership of the group.

Of course, culture does have an influence in sharpening or softening even those tendencies that are most deeply rooted in our human nature. And there can be variations between individuals. Nothing I have said is contradicted by the existence of individuals who do not care for their kin, or couples where the man looks after the children while the woman serves in the army. I must also stress that my rough classification of human behavior carries no evaluative overtones. I am not saying that because male dominance is characteristic of almost all societies, it is therefore good or acceptable, or we should not attempt to change it. My point is not about deducing an "ought" from an "is," but about estimating the price we may have to pay for achieving our goals.

For example, if we live in a society with a hierarchy based on a hereditary aristocracy, and we abolish that aristocracy, as the French and American revolutionaries did, we are likely to find that a new hierarchy emerges, based perhaps on military power or wealth. When the Bolshevik revolution in Russia abolished both the hereditary aristocracy and private wealth, a hierarchy soon developed on the basis of rank and influence within the Communist Party; this became the basis for all sorts of privileges. The tendency to form hierarchies shows itself in petty ways in corporations and bureaucracies, where people place enormous importance on how big their office is and how many windows it has. None of this shows that hierarchy is good, or desirable, or even inevitable; but it does show that getting rid of it is not going to be as easy as previous revolutionaries thought.

THE LEFT HAS TO ACCEPT and understand our nature as evolved beings. But there are different ways of working with the tendencies inherent in human nature. The market economy is based on the idea that human beings can be relied upon to work hard and show initiative only if by doing so they will further their own economic interests. To serve our own interests we will strive to produce better goods than our competitors, or to produce similar goods more cheaply. Thus, as Adam Smith said, the self-interested desires of a multitude of individuals are drawn together, as if by a hidden hand, to work for the benefit of all. Garrett Hardin summarized this view in *The Limits of Altruism* when he wrote that public policies should be based on "an unwavering adherence to the cardinal rule: never ask a person to act against his own self-interest." In theory—abstract theory, that is, free from any assumptions about human nature—a state monopoly should be able to provide the cheapest and most efficient utility services, transport services, or, for that matter, bread supply; indeed, such a monopoly would have huge advantages of scale and would not have to make profits for its owners. However, when we take into account the popular assumption that self-interest—more specifically, the desire to enrich oneself—drives human beings to work well, the picture changes. If the community owns an enterprise, its managers do not profit from its success. Their own economic interest and that of the enterprise pull in different directions. The result is, at best, inefficiency; at worst, widespread corruption and theft. Privatizing the enterprise will ensure that the owners will take steps to reward management in accordance with performance; in turn, the managers will take steps to ensure that the enterprise runs as efficiently as possible.

This is one way of tailoring our institutions to human nature, or at least to one view of human nature. But it is not the only way. Even within the terms of Hardin's cardinal rule, we still have to ask what we mean by "self-interest." The acquisition of material wealth, beyond a relatively modest level, has little to do with self-interest in the biological sense of maximizing the number of descendants one leaves in future generations. There is no reason to assume that increasing personal wealth must, either consciously or unconsciously, be the goal that people set for themselves. It is often said that money cannot buy happiness. This may be trite, but it implies that it is more in our interests to be happy than to be rich. Properly understood, self-interest is broader than economic self-interest. Most people want their lives to be happy, fulfilling, or meaningful; they recognize that money is at best a means of achieving part of these ends. Public policy does not have to rely on self-interest in this narrow economic sense.

Modern Darwinian thought embraces both competition and reciprocal altruism (a more technical term for cooperation). Focusing on the competitive element, modern market economies are premised on the idea that we are driven by acquisitive and competitive desires. Free market economies are designed to channel our acquisitive and competitive desires so that they work for the good of all. Undoubtedly, this is better than a situation in which they work only for the good of a few. But even when competitive consumer societies work at their best, they are not the only way of harmonizing our nature with the common good. Instead we should seek to encourage a broader sense of self-interest, in which we seek to build on the social and cooperative side of our nature rather than the individualistic and competitive side.

Robert Axelrod's work on the prisoner's dilemma provides a basis for building a more cooperative society. The prisoner's dilemma describes a situation in which two people can each choose whether or not to cooperate with the other. The catch is that each does better, individually, by not cooperating; but if both make this choice, they will both be worse off than they would have been if they had both chosen to cooperate. The outcome of rational, self-interested choices by two or more individuals can make all of them worse off than they would have been if they had not pursued their own interest. The individual pursuit of self-interest can be collectively self-defeating.

People who commute to work by car face this kind of situation every day. They would all be better off if, instead of sitting in heavy traffic, they abandoned their cars and used the buses, which would then travel swiftly down uncrowded roads. But it is not in the interests of any individual to switch to the bus, because as long as most people continue to drive, the buses will be even slower than cars.

Axelrod is interested in which strategy—cooperating or not cooperating—would bring about the best results for parties who face repeated situations of this type. Should they always cooperate? Should they always defect, as the noncooperative strategy suggests? Or should they adopt some mixed strategy, varying cooperation and defection in some way? Axelrod invited people to suggest strategies which would produce the best payoff for the person using it, if they were in repeated prisoner's dilemma situations.

When he received the answers, Axelrod ran them against each other on a computer, in a kind of round-robin tournament in which each strategy was pitted against every other strategy 200 times. The winner was a simple strategy called "tit for tat." It opened every encounter with a new player by cooperating. After that, it simply did whatever the other player had done the previous time. So if the other cooperated, it cooperated; and it continued to do so unless the other defected. Then it defected too, and continued to do so unless the other player again cooperated. Tit for tat also won a second tournament that Axelrod organized, even though the people sending in strategies this time knew that it had won the previous tournament.

Axelrod's results, which have been supported by subsequent work in the field, can serve as a basis for social planning that should appeal to the left. Anyone on the left should welcome the fact that the strategy with the best payoff always begins with a cooperative move and is never the first to abandon the cooperative strategy or seek to exploit the "niceness" of the other party. But members of the more idealistic left may regret that tit for tat does not continue to cooperate no matter what. A left that understands Darwin must realize that this is essential to its success. By being provokable, tit for tat creates a virtuous spiral in which life gets harder for cheats, and so there are fewer of them. In Richard Dawkins's terms, if there are "suckers," then there will also be "cheats" who can prosper by taking advantage of them. It is only by refusing to be played for a sucker that tit for tat can make it possible for cooperators to do better than cheats. A non-Darwinian left would blame the existence of cheats on poverty, or a lack of education, or the legacy of reactionary ways of thinking. A Darwinian left will realize that while all these factors may make a difference to the level of cheating, the only permanent solution is to change the payoffs so that cheats do not prosper.

The question we need to address is: under what conditions will tit for tat be a successful strategy for everyone to adopt? The first problem is one of scale. Tit for tat cannot work in a society of strangers who will never encounter each other again. This is why people living in big cities do not always show the consideration to each other that is the norm in rural villages in which people have known each other all their lives. We need to find structures that can overcome the anonymity of the huge, highly mobile societies in which we live, and which show every sign of increasing in size.

The next problem is even more difficult. If nothing you do really makes much difference to me, tit for tat will not work. So while equality is not required, too great a disparity in power or wealth will remove the incentive for mutual cooperation. If you leave a group of people so far outside the social commonwealth that they have nothing to contribute to it, you alienate them from the social practices and institutions of which they are part; and they will almost certainly become adversaries who pose a threat to those institutions. The political lesson of twentieth-century Darwinian thinking is entirely different from that of nineteenth-century social Darwinism. Social Darwinists saw the fact that those who are less fit will fall by the wayside as nature's way of weeding out the unfit—an inevitable result of the struggle for existence. To try to overcome it was futile, if not positively harmful. A Darwinian left which understands the prerequisites for mutual cooperation and its benefits would strive to avoid economic conditions that create outcasts.

LET ME DRAW some threads together. What distinguishes a Darwinian left from previous versions of the left? First, a Darwinian left would not deny the existence of a human nature or insist that human nature is inherently good or infinitely malleable. Second, it would not expect to end all conflict and strife between human beings. Third, it would not assume that all inequalities are due to discrimination, prejudice, oppression, or social conditioning. Some will be, but not all. For example, the fact that there are fewer women chief executives than men may be due to men's being more willing to subordinate their personal lives and other interests to their career goals; biological differences between men and women may be a factor in that greater readiness to sacrifice everything for the sake of getting to the top.

What about those things that a Darwinian left would support? First, it would recognize that there is such a thing as human nature. It would seek to find out more about it so that it can be grounded on the best available evidence of what human beings are. Second, it would expect that, under many different social and economic systems, many people will act competitively in order to enhance their own status, gain power, and advance their interests and those of their kin. Third, it would expect that irrespective of the social and economic system in which they live, most people will respond positively to invitations to enter into mutually beneficial forms of cooperation, as long as the invitations are genuine. Fourth, it would promote structures that foster cooperation rather than competition, and it would attempt to channel competition into socially desirable ends.

Fifth, it would recognize that the way in which we exploit nonhuman animals is a legacy of a pre-Darwinian past which exaggerated the gulf between humans and other animals, and therefore work toward a higher moral status for nonhuman animals. Sixth, it would stand by the traditional values of the left by being on the side of the weak, poor, and oppressed, but think very carefully about what will really work to benefit them.

In some ways, this is a sharply deflated vision of the left, its utopian ideas replaced by a coolly realistic view of what can be achieved. But in the longer term, we do not know to what extent our capacity to reason can take us beyond the conventional Darwinian constraints on the degree of altruism that a society may be able to foster. We are reasoning beings. Once we start reasoning, we may be compelled to follow a chain of argument to a conclusion that we did not anticipate. Reason provides us with the capacity to recognize that each of us is simply one being among others, all of whom have wants and needs that matter to them, as our needs and wants matter to us. Can this insight ever overcome the pull of other elements in our evolved nature that act against the idea of an impartial concern for all of our fellow humans, or better still, for all sentient beings?

No less a champion of Darwinian thought than Richard Dawkins holds out the prospect of "deliberately cultivating and nurturing pure, disinterested altruism—something that has no place in nature, something that has never existed before in the whole history of the world." Although "we are built as gene machines," he tells us, "we have the power to turn against our creators." There is an important truth here. We are the first generation to understand not only that we have evolved, but also the mechanisms by which we have evolved. In his philosophical epic, *The Phenomenology of Mind*, Hegel portrayed the culmination of history as a state of absolute knowledge, in which the mind knows itself for what it is, and thus achieves its own freedom. We don't have to accept Hegel's metaphysics to see that something similar really has happened in the last fifty years. For the first time since life emerged from the primeval soup, there are beings who understand how they have come to be what they are. In a more distant future we can still barely glimpse, it may turn out to be the prerequisite for a new kind of freedom: the freedom to shape our genes so that instead of living in societies constrained by our evolutionary origins, we can build the kind of society we judge best.

A Meaningful Life

FROM *Ethics into Action: Henry Spira and the Animal Rights Movement*

TO SAY that life is essentially meaningless is to express an attitude, not to state a fact. For that reason—and unlike the assumption that an individual cannot make a difference to the world—it is not an assertion that can be refuted simply by pointing to facts about Henry Spira's life. But if, when we face the end of our life, we can look back on it with the satisfaction and fulfillment that come from believing that we have spent our life doing something that was both worthwhile and interesting to do, then perhaps that is enough to show that we have found a way to make life meaningful. That has been Henry's experience.

The best way in which I can describe how Henry has found his life meaningful is to explain how this book came to be written. I can't recall exactly when I first told Henry that I would like to write a book about him, but the idea had been with me for many years. One sunny October day in 1992, we walked into Central Park, found a lawn with a view of the midtown skyline, and made ourselves comfortable on the grass. I pulled out a tape recorder and for an hour or two asked Henry questions about his life. I left him with the tape, which he said he would get typed up. Then I returned home to Melbourne, and my time was immediately swallowed up by other work. Something similar must have happened to Henry, because for a long time no typescript of the interview arrived. Given my other commitments, I was relieved that Henry, instead of urging me to get on with the promised biography, had himself apparently let it sink down his priority list.

The typescript finally arrived in 1994, but I was still too busy with other work to do anything with it. In 1995, I was selected by the Australian Greens to lead their senate ticket for my home state, Victoria, in the next federal election. When I saw Henry that year, he must have been thinking about his own mortality—he was sixty-seven then—because he asked me if I was still thinking of doing the book and, if so, what instructions I wanted put in his will about what should be done with the papers that, systematically filed and shelved, filled every room in his apartment from floor to ceiling. I said that in principle I was still interested, but if I were elected to the senate, I wouldn't be able to do anything about it during my term of office, which would last six years. On the other hand, if I were not elected, I said, there was a good chance that I could find some time to work on the book rather soon.

The election was held in March 1996, and I was not elected. No doubt to remind myself that this disappointing result did have its positive side, I sketched out an itinerary for an overseas trip, built around invitations to speak in Europe in May and at a March for the Animals in Washington at the end of June. On April 21, I sent Henry a fax telling him that since I had not been elected, "I'm starting to think about what else to do with the rest of my life. The book about you is one possibility, some time in the next two or three years." Could I stay with him for a few days in June, before my Washington commitment, so that we could talk about it?

That evening I had a message on my answering machine. It was unmistakably Henry's voice, saying that he wanted to speak to me and would call again soon, but there was something very troubling about his tone. I reached for the phone to call him back, but before I could do so, it rang again.

"Peter?"

"Henry, how are you?" I asked.

"Lousy, actually."

"Why, what's the matter?"

"I've got an adenocarcinoma of the esophagus, grade three."

"What does that mean, in layman's terms?"

"Let me put it this way: If you could choose the kind of cancer you were going to have, you wouldn't choose this one."

I made some inadequate kind of response. Henry then said that while he'd really like me to do the book, he wasn't sure that he was still going to be around in late June.

I was in New York six days later. Over the next five days, I slept on the sofa bed in Henry's apartment, and we spent all our waking time together. Henry had lost a lot of weight and lacked the energy I was used to seeing in him. He had to be pushed hard to tell me about his illness, but eventually I learned that for years he had occasionally had to vomit after eating. In 1995, the problem had become worse. In September, he had had a barium examination. It revealed a suspicious obstruction in his esophagus. Henry had never concerned himself much with his own health, and for a time he tried to put off doing anything about it. By February, he finally had to accept that he could put it off no longer. On March 4, he was admitted to New York University Medical Center and was operated upon. The operation found a tumor in his esophagus. The surgeon cut out a large part of the esophagus and adjoining areas of his stomach. Henry spent the next ten days in the hospital before he was able to go home. Now, seven weeks after the operation, he was still weak and had trouble keeping any food down. The outlook was even worse: the cancer was invasive, and the pathology report showed that it had spread into some of his lymph nodes. His life expectancy was a matter of months. His doctor had recommended radiation and chemotherapy, but he was unable to give Henry any statistics to show that this would help. Henry checked out the literature himself and found that there was no evidence that radiation or chemotherapy offered any significant life-prolonging effect for the kind of cancer he had. What he did know was that it would make him feel very bad. He rejected his doctor's recommendation. That wasn't the only recommendation Henry rejected. His friends and acquaintances suggested an amazing number of unorthodox cures for cancer, ranging from special diets to having all his fillings removed. He didn't try any of them. Instead, he began looking for a doctor who would help him die when he had had enough. Meanwhile, there was work to do.

During my time in New York, Henry and I worked hard on making this book possible. Before leaving Melbourne, I had reasoned that if Henry wasn't going to be around much longer, it would be a good idea to record some interviews on videotape. I didn't have any specific idea about what I might do with the tapes, but I wanted as much as possible of the Henry I knew to be preserved for posterity—not just the words he said, but the way he said them. So at my suggestion, Henry phoned Julie Akeret, an independent filmmaker who had once made a short film about me called *In Defense of Animals.* Julie came over with a cameraman she knew, and despite Henry's weak condition, we taped several hours of interviews, which provided the outline for this book and many of the quotations used in it.[1]

Henry gave me the contact details for many people who had been important in his life. I called some of them from his apartment. Many—including his sister Renée, who lived only an hour away, on Long Island—had not been in touch with Henry for some time and had no idea how ill he was. Henry had not tried to hide the news, but he hadn't felt like phoning and saying, "Hey, I've got cancer and will probably die in a month or two."

The most remarkable thing about Henry during this period was the total absence of any sign of depression. Life had been good, he said; he had done what he wanted to do and had enjoyed it a lot. Why should he be depressed? The thing that really worried him about the cancer was that he would die a slow, lingering death. He was looking for a doctor who would help him to die sooner rather than later, and at home rather than in a hospital, where he feared losing control over his own life. While I was staying with him, he went to a doctor and came back to the apartment with a bottle of pills that the doctor had given him—officially for pain relief. Together we looked up the drug in a pharmacopoeia that Henry had. The bottle contained about four times the lethal dose. Henry's relief was palpable. With that worry taken care of, he seemed remarkably untroubled by the fact that he was expecting to die soon.

Henry did not die in the time his doctors predicted. When I returned to New York in June, on my way back from Europe for the March for the Animals, he was markedly stronger than he had been at the end of April. He even went to Washington and spoke at the march, though he had always been a bit cynical about the value of spending too much energy on activities that lacked a specific goal. As this book goes to press, in March 1998, Henry is still very much alive and working hard on farm animal issues, targeting the fast-food chains McDonald's, KFC, and Burger King. He is also watching the development of the Center for a Livable Future. I can't help wondering if his strong sense that the biggest gains for animals still lie ahead has kept him going far longer than the nature of his cancer gave him any right to expect.

One mark of living well is to live so that you can accept death and feel satisfied with what you have done with your life. Henry's life has lacked many of the things that most of us take for granted as essential to a good life. He has never married or had a long-term, live-in relationship. He has no children. His father and one of his sisters committed suicide, and his mother was mentally ill for much of her life. His relationship with Renée, the sole surviving member of his immediate family, is not close. His rent-controlled apartment, while spacious and well situated, is spartan. He doesn't go to movies, to concerts, to the theater, or to fine restaurants. He hasn't taken a vacation for twenty years. Yet at the age of sixty-eight he was able to contemplate his own imminent death with no major regrets about the way he had lived. What makes up for the absence of so much that, for most people, are the essentials of a good life?

In our 1992 interview, I tried to locate the source of Henry's satisfaction:

PETER: So, looking back on what you've been doing for the last twenty to thirty years, what do you feel about it? What sort of a life has it been?

HENRY: Well, I think for one thing, I've totally enjoyed it. And I think that if I had a choice of what it is that I wanted to do, that's what I would have wanted to do. And looking back, I think it was worth the effort, it was worth the energy, and I think that I pushed things along, as best I could.

PETER: Some people might say that you've sacrificed a lot of time and effort, while not doing very much for yourself.

HENRY: I've never felt that I've sacrificed for others. I just felt that I'm doing what I really want to do and what I want to do most. And I feel most alive when I'm doing it.

PETER: Is that a matter of personal temperament? What's the secret of why you enjoy it?

HENRY: I don't know why I enjoy it, but I think one can be a lot more effective if one really feels good about doing it, if one gets up in the morning just raring to pick up where one left off the night before—as opposed to doing it for others, doing it because it should be done, doing it because it's the right thing to do.

PETER: What if someone said that you get your kicks out of sticking it into other people, like Frank Perdue? (A chicken mogul whom Henry attacked in one of his campaigns.)

HENRY: I don't think I've ever stuck it into others just for the sake of sticking it in. I think we try to dialogue. I think the pleasure you really get is out of conceptualizing a campaign and moving it along. And you want to move it along the fastest way possible. The fastest way possible is to move it along with cooperation and collaboration. It's only when you get forced into an adversarial position that you then try to do the best you can in that direction.

Peter: You couldn't say that you really minded being in an adversarial position, could you?

Henry: No, I think once you're in it, one sort of thrives on it. But it's basically that one is forced into it to begin with. I think I'm comfortable working either way; but I think the thing that you really get your satisfaction out of is conceptualizing a campaign that you know is absolutely going to work, and then seeing it work.

The real satisfaction, Henry told me on another occasion, is "not the fact that you made somebody else feel like gone-over garbage," but the "creative high" that comes from getting all the pieces of the puzzle together, which gives him a sense that "lightning has struck."

The idea that no matter how serious the cause for which you are working, you should still enjoy what you are doing is one that Henry has held for a long time. Among the radical thinkers he had read in his youth was the American anarchist Emma Goldman. Goldman liked dancing, a pastime that her more puritanical anarchist friends regarded as frivolous. It was not, they told her, an activity fit for a revolutionary. Goldman responded: "If I can't dance, I don't want your revolution." It was a line that had always struck a chord with Henry:

> The point [Goldman] is making is, you've got to enjoy what you're doing to be effective. What you're doing is what you've absolutely got to be doing, not because you feel you've got to do it but, rather, because this is what your life is about. You feel good when you're doing it. . . . I feel best when I'm doing something that's going to make a difference. When I go, I want to look back and say, "I made this a better place for others." But it's not a sense of duty, rather this is what I want to do. . . . I feel best when I'm doing it well.

As for the more common idea that you enjoy yourself best by earning a lot of money and living it up, Henry rejects that position: "When I was working on the ships I had so much money I didn't know where to put it. I stayed in some of the best places. . . . It was interesting for the experience, but I didn't want the lifestyle. It didn't give me a high."

Although Henry emphasizes that he has chosen his life because he feels good about what he does, rather than because some sense of duty makes him feel that it is the right thing to do, there is no doubt that he is motivated by a strong sense of doing something worthwhile:

> I guess basically one wants to feel that one's life has amounted to more than just consuming products and generating garbage. I think that one likes to look back and say that one's done the best one can to make this a better place for others. You can look at it from this point of view: What greater motivation can there be than doing whatever one possibly can to reduce pain and suffering?

While others may feel the same motivation, few manage to keep it going throughout an entire life. In a magazine interview in 1995, Henry was asked whether, in view of the size of the problems he is tackling, he ever gets tired of trying. He replied:

> It's crucial to have a long-term perspective. Looking back over the past twenty years, I see progress that we've helped achieve. And when a particular initiative causes much frustration, I keep looking at the big picture while pushing obstacles out of the way. And there's nothing more energizing than making a difference.[2]

During my visit in April 1996, when Henry and I thought his life nearly over, I asked him to sum up what he thought he had achieved. He said:

> I've pushed the idea that activism has to be results-oriented, that you can win victories, that you can fight city hall, and that if you don't like to be pushed around and you don't like to see others pushed around, you can have an impact. . . . It's like this guy from the *New York Times* asked me what I'd like my epitaph to be. I said, "He pushed the peanut forward." I try to move things on a little.[3]

I asked if he was satisfied with having achieved that.

> I might have done some things differently, but on the whole, I've given it the best shot that I've got. . . . Looking back on my life, it's been satisfying. I've done a lot of things I wanted to do. I've had an enormous amount of fun doing it, and if I were going to do it over again, I'd do it very similar to the way I have done it.

Henry Spira died peacefully in December 1998.

Autobiographical Notes

Animal Liberation: A Personal View

FROM *Between the Species*

I ARRIVED IN OXFORD in October 1969. I had come to do a graduate degree in philosophy—the natural climax to the education of an Australian philosophy student preparing for an academic career. My interests were in ethics and political philosophy, but the connections between my philosophical studies and my everyday life would have been hard to discern. My day-to-day existence and my ethical beliefs were much like those of other students. I had no distinctive views about animals or about the ethics of our treatment of them. Like most people, I disapproved of cruelty to animals, but I was not greatly concerned about it. I assumed that the R.S.P.C.A. and the government could be relied upon to see that cruelty to animals was an isolated occurrence. I thought of vegetarians as, at best, otherworldly idealists and, at worst, cranks. Animal welfare I regarded as a cause for kindly old ladies rather than for serious political reformers.

The crack in my complacency about our relations with animals began in 1970 when I met Richard Keshen. Our meeting was entirely accidental. Richard, a Canadian, was also a graduate student in philosophy. He and I were attending lectures given by Jonathan Glover on free will, determinism, and moral responsibility. They were stimulating lectures, and when they finished a few students often remained behind to ask questions or to discuss points with the lecturer. After one particular lecture, Richard and I were among this small group; when Glover had answered our queries, we walked out together, discussing the issue further. It was lunchtime, and Richard suggested that we go to his college, Balliol, and continue our conversation over lunch. When it came to selecting our meal, I noticed that Richard asked if the spaghetti sauce had meat in it and, when told that it had, took a meatless salad. So, when we had talked enough about free will and determinism, I asked Richard why he had avoided meat. That began a discussion that was to change my life.

The change did not take place immediately. What Richard Keshen told me about the treatment of farm animals, combined with his arguments against our neglect of the interests of animals, gave me a lot to think about, but I was not about to change my diet overnight. Over the next two months, together with my wife, Renata, I met Richard's wife, Mary, and the two other Canadian philosophy students, Roslind and Stanley Godlovitch, who had been responsible for Richard's and Mary's becoming vegetarians. Ros and Stan had become vegetarians a year or two earlier, before reaching Oxford. They had come to see our treatment of nonhuman animals as analagous to the brutal exploitation of other races by whites in earlier centuries. This analogy they now urged on us, challenging us to find a morally relevant distinction between humans and nonhumans which could justify the differences we make in our treatment of those who belong to our own species and those who do not.

During these two months, Renata and I read Ruth Harrison's pioneering attack on factory farming, *Animal Machines*. I also read an article which Ros Godlovitch had recently published in the academic journal *Philosophy*. She was in the process of revising it for republication in a book that she, Stan, and John Harris, another vegetarian philosophy student at Oxford, were editing. Ros was a little unsure about the revisions she was making, and I spent a lot of time trying to help her clarify and strengthen her arguments. In the end, she went her own way, and I don't think any of my suggestions were incorporated into the revised version of the article as it appeared in *Animals, Men and Morals*—but in the process of putting her arguments in their strongest possible form, I had convinced myself that the logic of the vegetarian position was irrefutable. Renata and I decided that if we were to retain our self-respect and continue to take moral issues seriously, we should cease to eat animals.

Through the Keshens and the Godlovitches, we got to know other members of a loose group of vegetarians. Several of them lived together in a rambling old house with a huge vegetable garden. Among the residents of this semicommunal establishment were John Harris and two other contributors to *Animals, Men and Morals,* David Wood and Michael Peters. Philosophically, we agreed on little but the immorality of our present treatment of animals. David Wood was interested in continental philosophy, Michael Peters in Marxism and structuralism; Richard Keshen's favorite philosopher was Spinoza; Ros Godlovitch was still developing her basic position—she had not studied philosophy as an undergraduate and became involved in it only as a result of her interest in the ethics of our relations with animals—and Stan Godlovitch refused to work on moral philosophy, restricting himself to the philosophy of biology. I was more in the mainstream of Anglo-American philosophy than any of the others, and in moral philosophy I took a much more utilitarian line than they did.

Also around Oxford at this time were Richard Ryder, Andrew Linzey, and Stephen Clark. Richard Ryder was working at the Warneford Hospital in Oxford. He had written a leaflet on "Speciesism"—the first use of the term, as far as I know—and now was writing an essay on animal experimentation for *Animals, Men and Morals.* Later, he developed this work into his splendid attack on animal experimentation, *Victims of Science.* He was also organizing a "ginger group" within the R.S.P.C.A., with the aim of getting that then extremely conservative body to eject its foxhunters and take a stronger stance on other issues. That seemed a very long shot, then. I was introduced to Richard Ryder through Ros Godlovitch, and from him I learned a lot about animal experimentation. At the time, our positions were the mirror image of each other—I was a vegetarian but not a strong opponent of animal experimentation, because I naively thought most experiments were necessary to save lives and were, therefore, justified on utilitarian grounds. Richard Ryder, on the other hand, was not then a vegetarian but was opposed to animal experimentation, because of the extreme suffering it often involved.

Andrew Linzey was interested in the animal issue from the point of view of Christian theology, which was not the concern of most of the group, for we were a nonreligious lot. His book, *Animal Rights,* was published by the SCM Press in 1976. Stephen Clark was a Fellow of All Souls College, Oxford, during this period, but I did not get to know him until much later, after he had written *The Moral Status of Animals,* which appeared in 1977.

Animals, Men and Morals, the first of all these books, appeared in 1971. We had great hopes for it, for it demanded revolutionary change in our attitudes toward and treatment of nonhuman animals. I think Ros Godlovitch, especially, thought that the book might trigger a widespread protest movement. In the light of these expectations, the book's reception was profoundly disappointing. The major newspapers and weeklies ignored it. In the *Sunday Times*, for example, it was mentioned only in the "In Brief" column—just one short paragraph of exposition, without a comment. Our ideas seemed to be too radical to be taken seriously by the staid British press.

At the time, the virtual silence which met the British publication of *Animals, Men and Morals* seemed a severe setback. Yet it turned out to be the first of a chain of events that led me to write *Animal Liberation*. Some time after *Animals, Men and Morals* appeared in England, the Godlovitches got some better news: Taplinger had agreed to publish an American edition. But would the book get more attention in America than in Britain? I determined to do my best to see that it did. I had, in any case, been wanting to write something to make people more aware of the injustice of our treatment of animals but had been deterred from doing so by the feeling that since so many of my ideas had come from others, and especially from Ros, I should allow her to publish them. Now, I thought of a way to satisfy my own desire to do something to make people aware of the issue while at the same time helping to get my friends' ideas the attention they deserved but had not received. I would write a long review-article, based on *Animals, Men and Morals* but drawing the views of the several contributors together into a single coherent philosophy of animal liberation. There was only one place I knew of in America where such a review-article might appear: *The New York Review of Books*.

I wrote to the editors of the *New York Review*, describing the book and the review I would write. I did not know what answer to expect, since I had had no previous contact with them, and they would never have heard of me. I knew that they were open to novel and radical ideas, but did they, perhaps, accept contributions only from people they knew? Would the idea of animal liberation seem ridiculous to them?

Robert Silvers's reply was guardedly encouraging. The idea was intriguing, and he would like to see the article, though he could not promise to publish it. That was all the encouragement I needed, however, and the article was soon written and accepted. Entitled "Animal Liberation," it appeared in April 1973. I was soon receiving enthusiastic letters from people who seemed to have been waiting for their feelings about the mistreatment of animals to be given a coherent philosophical backing.

Among the letters was one from a leading New York publisher, suggesting that I develop the ideas sketched in the article into a full-length book. Although my review had helped *Animals, Men and Morals* become better known in America—it eventually went into a paperback edition there, something that never happened in Britain—there was obviously room for a different kind of book, more systematic in its approach than a compilation of articles by different authors can be. There was also a need for factual research to be done on factory farming and experimentation in America, since the data in both *Animal Machines* and *Animals, Men and Morals* were largely British. By this time, I knew that I would soon be leaving Oxford, for I had accepted a visiting position at New York University, which would make a good base for this kind of research. So, during our last summer in Oxford, I began work on *Animal Liberation*.

I arrived in New York in September 1973, with an outline of the book and the drafts of what were to become the first chapter—which contains the core of the ethical argument—and the historical chapter, which outlines the development of speciesism in Western thought. But there had been a change of personnel at the New York publisher that had suggested the book, and they were no longer interested. Fortunately, Robert Silvers suggested that the *New York Review* might publish the book—it had done one or two others which had grown out of articles it had first published, and Silvers was personally very supportive of the idea of animal liberation. While teaching at New York University, I spent as much time as I could on the book, and when my visiting appointment ended in June, I began to work on it full time, spending many of my days either at the New York Public Library reading farming magazines, or at the offices of United Action for Animals going through the extensive files on animal experimentation, which the late Eleanor Seiling was happy to make available to me. The book was finished by Christmas, just before we left New York to return to Australia.

Animal Liberation was not an immediate success. It got some good reviews, as well as some silly ones, and it sold steadily but not spectacularly. Some of the leading philosophical journals devoted special issues to the topic, which was gratifying. Nevertheless, my highest hopes—naive ones, perhaps—were not fulfilled. The book did not spark an immediate upsurge against factory farming and animal experimentation. The only prompt political effect was that Congressman Ed Koch, as he then was, referred to the book in a speech in which he proposed a National Commission of Inquiry into animal welfare matters. But his proposal lapsed, and for a few years it seemed that the only effect the book was having was to put the question of animals and ethics onto the list of topics discussed in applied ethics courses in university philosophy departments.

At the time, it was impossible to detect—at least from my distant vantage point in Australia—how the animal liberation movement was gradually gathering force in Britain and the United States. It is really only in the past five years that the animal liberation movement has made real gains. I am not sure why this is so, except that there has been a lot of hard work by some very dedicated people. Five years ago, the public in most developed countries was largely unaware of the nature of modern intensive animal rearing. Now, in Britain, in West Germany, in Scandanavia, in the Netherlands, and in Australia, a large body of informed opinion is opposed to the confinement of laying hens in small wire cages, and of pigs and calves in stalls so small that they cannot walk a single step or even turn around. In Britain, a House of Commons Agriculture Committee has recommended that cages for laying hens be phased out. Switzerland has gone one better, actually passing legislation which will get rid of the cages by 1992. A West German court pronounced the cage system contrary to the country's anticruelty legislation, and although the government found a way of rendering the court's verdict ineffective, the West German state of Hesse recently announced that it would follow Switzerland's example and begin to phase the cages out. Perhaps the most positive step forward for British farm animals has been in the worst of all forms of factory farming, the so-called "white veal trade." Veal calves were standardly kept in darkness for 22 hours a day, in individual stalls too small for them to turn around. They had no straw to lie on—for fear that by chewing it, they would cause their flesh to lose its pale softness—and were fed on a diet deliberately made deficient in iron, so that the flesh would remain pale and fetch the highest possible price in the gourmet restaurant trade. A campaign against the trade led to a widespread consumer boycott. As a result, Britain's largest veal producer conceded the need for change and moved its calves out of their bare, wooden, five-feet-by-two-feet stalls into group pens with room to move and straw for bedding.

There have also been important gains in the area of animal experimentation. In contrast to the situation with factory farming, these have occurred mostly in the United States. The first success came in 1976, in a campaign against the American Museum of Natural History. The museum was selected as a target because it was conducting a particularly pointless series of experiments which involved mutilating cats to investigate the effect this had on their sex lives. In June 1976, animal liberation activists began picketing the museum, writing letters, advertising, and gathering support. They kept it up until, in December 1977, it was announced that the experiments would no longer be funded.

This victory may have saved no more than sixty cats from painful experimentation, but it had shown that a well-planned, well-run campaign can prevent scientists from doing as they please with laboratory animals. Henry Spira, the New York ex-merchant seaman, ex-civil rights activist who had led the campaign against the museum, used the victory as a stepping-stone to bigger campaigns. He now runs two coalitions of animal groups, focusing on the rabbit-blinding Draize eye test and on the LD50, a crude, fifty-year-old toxicity test designed to find the lethal dose for 50 percent of a sample of animals. Together, these tests inflict suffering and distress on more than five million animals yearly in the United States alone.

Already the coalitions have begun to reduce both the number of animals used and the severity of their suffering. United States government agencies have responded to the campaign against the Draize test by moving to curb some of the most blatant cruelties. They declared that substances known to be caustic irritants, such as lye, ammonia, and oven cleaners, need not be retested on the eyes of conscious rabbits. If this seems too obvious to need saying by a government agency, that merely indicates how bad things were until the campaign began. The agencies have also reduced by one-half to one-third the suggested number of rabbits needed per test for other products. Two major companies—Procter and Gamble and Smith, Kline, and French—have released programs for improving their toxicology tests which should involve substantially less suffering for animals. Another company, Avon, reported a decline of 33 percent in the number of animals it uses.

In the most recent, and potentially most significant, breakthrough, the United States Food and Drug Administration has announced that it does not require the LD50. At a stroke, corporations developing new products have been deprived of their standard excuse for using the LD50—the claim that the FDA forces them to do the test if the products are to be released onto the American market.

The same five years that have seen the gains I have mentioned have also seen a steady rise in militancy in the movement. In Britain, Canada, France, West Germany, Italy, and Australia, animals have been released and laboratories have been damaged. The U.S. Animal Liberation Front gained national publicity in May 1984, when its members entered the laboratory of Dr. Thomas Gennarelli at the University of Pennsylvania, damaged equipment, and took a number of videotapes. The videotapes, sections of which were subsequently shown on television, show severe head injuries being inflicted on monkeys. Injured monkeys, their limbs flapping uncontrollably, are tied to chairs while experimenters try to get them to respond. On the videotapes, the research team jokes about the monkeys' injuries, and Dr. Genarelli refers to the animals as "suckers." In July 1985, the United States announced that its own inquiry into the laboratory had shown serious deficiencies in animal care, and funding for the laboratory was suspended. Another raid at the City of Hope research laboratory in Duarte, California, also led to an inquiry which stopped experiments at the laboratory.

The research community is understandably alarmed. Many laboratories have increased their security arrangement, but this is a costly business, and money spent on fences and guards is presumably not then available for research—which is just what the animal liberation activists want. To guard every factory farm would be even more expensive. No wonder that some of those who experiment on animals or raise them for food hope that animal liberation will just prove to be a passing fad.

That hope is bound to be disappointed. The animal liberation movement is here to stay. It has been building steadily now for more than a decade. There is wide public support for the view that we are not justified in treating animals as mere things to be used for whatever purposes we find convenient, whether it be the entertainment of the hunt or as a laboratory tool for the testing of some new food coloring.

But there is still the question of the course the movement will take. Within the animal liberation movement, some forms of direct action have widespread support. Provided there is no violence against any animal, human or nonhuman, many activists believe that releasing animals from terrible situations and finding good homes for them is justified. They liken it to the illegal underground railroad which assisted black slaves to make their way to freedom. It is, they say, the only possible means of helping the victims of oppression.

In the worst cases of indefensible experiments, this argument is strong, but there is another question that should be asked by everyone interested not only in the immediate release of ten, or fifty, or a hundred animals, but in the prospects of a change that affects millions of animals. Is direct action effective as a tactic? Does it simply polarize the debate and harden the opposition to reform? So far, one would have to say, the publicity gained—and the evident public sympathy with the animals released—has done the movement more good than harm. This is, in large part, because the targets of these operations have been so well selected that the experimentation revealed is particularly difficult to defend.

Recently, this crucial matter of selecting only the most blatantly indefensible targets has not always been as strictly observed as it should have been. The groundswell of militant activity has increased, and some activists have gone beyond actions directed at releasing animals or documenting cruelty. In 1982, a group calling itself the "Animal Rights Militia" sent letter-bombs to Margaret Thatcher. The group had never been heard of before, has never been heard of since, and may not have been a genuine animal rights organization at all. But that is, of course, exactly the kind of activity which must be avoided, not only because it is disastrous as a tactic, but also because it is ethically indefensible in itself.

There are circumstances in which, even in a democracy, it is morally right to disobey the law, and the issue of animal liberation provides good examples of such circumstances. If the democratic process is not functioning properly, if repeated opinion polls confirm that an overwhelming majority opposes many types of experimentation, and yet the government takes no effective action to stop them, if the public is kept largely unaware of what is happening in factory farms and laboratories, then illegal actions may be the only available avenue for assisting animals and obtaining evidence about what is happening. My concern is not with breaking the law as such; it is with the prospect of the confrontation's becoming violent and leading to a climate of polarization in which reasoning becomes impossible and the animals themselves end up being the victims. Polarization between animal liberation activists, on the one hand, and the factory farmers and at least some of the animal experimenters, on the other hand, may be unavoidable. But actions which involve the general public or violent actions which lead to people's getting hurt would antagonize the community as a whole.

It is vital that the animal liberation movement avoid the vicious spiral of violence. Animal liberation activists must set themselves irrevocably against the use of violence, even when their opponents use violence against them. It is easy to believe that because some experimenters make animals suffer, it is all right to make the experimenters suffer. This attitude is mistaken. We may be convinced that people who abuse animals are totally callous and insensitive, but we lower ourselves to their level and put ourselves in the wrong if we harm or threaten to harm them. The entire animal liberation movement is based on the strength of its ethical concern. It must not abandon the high moral ground.

Instead of going down the same blind alley of violence and counter-violence, the animal liberation movement should follow the examples of the two greatest—and, not coincidentally, most successful—leaders of liberation movements in modern times: Gandhi and Martin Luther King. With immense courage and resolution, they stuck to the principle of non-violence despite the provocations and often violent attacks of their opponents. In the end, they succeeded because the justice of their cause could not be denied and because their behavior touched the consciences even of those who had opposed them. The struggle to extend the sphere of moral concern to nonhuman animals may be even harder and longer, but if it is pursued with the same determination and moral resolve, it will surely also succeed.

On Being Silenced in Germany

FROM *The New York Review of Books*

SOME SCENES from academic life in Germany and Austria today:

For the 1989/1990 winter semester, Dr. Hartmut Kliemt, a professor of philosophy at the University of Duisburg, a small town in the north of Germany, offered a course in which my book *Practical Ethics* was the principal text assigned to the class. First published in English in 1979, this book has been widely used in philosophy courses in North America, the United Kingdom, and Australia and has been translated into German, Italian, Spanish, and Swedish.[1] Until Kliemt announced his course, it had never evoked anything more than lively discussion. Kliemt's course, however, was subjected to organized and repeated disruption by protesters objecting to the use of the book on the grounds that in one of its ten chapters it advocates active euthanasia for severely disabled newborn infants. When after several weeks the disruptions showed no sign of abating, Kliemt was compelled to abandon the course.

The European society for the Philosophy of Medicine and Health Care is a learned society that does just what one would expect an organization with that name to do: it promotes the study of the philosophy of medicine and health care. In 1990 it planned its fourth annual conference, to be held in Bochum, Germany, in June. The intended theme of the conference was Consensus Formation and Moral Judgment in Health

Care. During the days leading up to the conference, literature was distributed in Bochum and elsewhere in Germany by the "Anti-Euthanasia Forum," stating that "under the cover of tolerance and the cry of democracy and liberalism, extermination strategies will be discussed. On these grounds we will attempt to prevent the Bochum Congress from taking place." On June 5, scholars who were about to attend the conference received a letter from the secretary of the society notifying them that it was being moved to Maastricht, in the Netherlands, because the German organizers (two professors from the Center for Medical Ethics at the Ruhr University in Bochum) had been confronted with "antibioethics agitation, threats and intimidation," and could not guarantee the safety of the participants.

In October 1990, Dr. Helga Kuhse, senior research fellow at the Centre for Human Bioethics at Monash University in Australia and author of *The Sanctity-of-Life Doctrine in Medicine: A Critique*,[2] was invited to give a lecture at the Institute for Anatomy of the University of Vienna. A group calling itself the "Forum of Groups for the Crippled and Disabled" announced that it would protest against the lecture, stating that "academic freedom has ethical limits, and we expect the medical faculty to declare that human life is inviolable." The lecture was then canceled by the faculty of medicine. The dean of the faculty, referring to Dr. Kuhse, told the press, "We didn't know at all who that was."[3]

The Institute for Philosophy at the University of Hamburg decided, with the agreement of faculty members and a student representative, to appoint a professor in the field of applied ethics. The list of candidates was narrowed down to six. At this point in selecting a professor in Germany, the standard procedure is to invite each of the candidates to give a lecture. The lectures were announced but did not take place. Students and protesters from outside the university objected to the advertising of a chair in applied ethics on the grounds that this field raised questions about whether some human lives were worth living. The protesters blocked the entrances to the lecture theaters and blew whistles to drown out any attempts by the speakers to lecture. The university canceled the lectures. A few weeks later, a new list of candidates was announced. Two philosophers active in the field of applied ethics were no longer in consideration; they were replaced by philosophers who have done relatively little work in applied ethics; one, for example, is best known for his work in aesthetics. One of those dropped from the short list was Dr. Anton Leist, author of a book that offers ethical arguments in defense of the right to abortion,[4] and also a coeditor of *Analyse & Kritik*, one of the few German journals publishing philosophy in the mode practiced in English-speaking countries. Ironically, a recent special issue of the journal was devoted to *Practical Ethics* and the issue of academic freedom in Germany.[5]

In February 1991 a roundtable discussion was to be held in Frankfurt, organized jointly by the adult education sections of both the Protestant and the Roman Catholic church. The theme was Aid in Dying, and among the participants was Norbert Hoerster, a highly respected German professor of jurisprudence, who has written in support of the principle of euthanasia. As the meeting was about to get under way, a group of people challenged the organizers, accusing them of giving a platform to a "fascist" and an "advocate of modern mass extermination." They distributed leaflets headed "No Discussion about Life and Death." The meeting had to be abandoned.

The International Wittgenstein Symposium, held annually at Kirchberg, in Austria, has established itself as one of the principal philosophical conferences on the continent of Europe. The fifteenth International Wittgenstein Conference was to have been held in August 1991, on the theme Applied Ethics. Arrangements for the program were made by philosophers from the Institute for Philosophy at the University of Salzburg. Among those invited to speak were Professor Georg Meggle, of the University of Saarbrücken; Professor R. M. Hare, former White's Professor of Moral Philosophy at the University of Oxford, and now a professor of philosophy at the University of Florida, Gainesville; and myself. When the names of those invited became known, threats were made to the president of the Austrian Ludwig Wittgenstein Society, Dr. Adolf Hübner, that the symposium would be disrupted unless the invitations to Professor Meggle and me were withdrawn. In other public discussions with opponents of the program, the threat of a boycott was extended to include several other invited professors: Hare, Kliemt, Hoerster, and Professor Dietrich Birnbacher of the department of philosophy at the Gesamthochschule in Essen.[6]

Dr. Hübner is not a philosopher; he is a retired agricultural veterinarian, so he read *Practical Ethics* only after the protest arose. On reading it, however, he formed the opinion that—as he wrote in an Austrian newspaper—the protests were "entirely justified."[7] In a long letter to the board of directors of the Austrian Ludwig Wittgenstein Society he wrote that "as a result of the invitations to philosophers who hold the view that ethics can be grounded and carried out in the manner of an objective critical science, an existential crisis has arisen for the Austrian Wittgenstein Symposium and the Wittgenstein Society."[8] The reference to the "objective critical science" is striking, since Hare, in particular, has devoted much of his life to insisting on the differences between ethical judgments and statements to which notions of objective truth or falsity are standardly applied.

According to some reports, opposition groups threatened to stage a display on "Kirchberg under the Nazis" if the invitations were not withdrawn. This threat proved so potent that innkeepers in Kirchberg were said to have stated that they would refuse to serve philosophers during the symposium.[9] To its considerable credit, the organizing committee resisted Dr. Hübner's proposal to withdraw the invitations from those philosophers against whom the protests were directed. Instead, it recommended that the entire symposium be canceled, since Dr. Hübner's public intervention in the debate had made it unlikely that it could be held without disruption. This recommendation was accepted by the committee of the Austrian Wittgenstein Society, against the will of Dr. Hübner himself. There will be no Wittgenstein Symposium in 1991.

For those who believe that there is a strong consensus throughout Western Europe supporting freedom of thought and discussion in general, and academic freedom in particular, these scenes come as a shock. How they have come about, however, is not so difficult to explain. The story has its beginnings in events in which I was directly involved. It stems from an invitation I received to speak, in June 1989, at a European Symposium on Bioengineering, Ethics, and Mental Disability, organized jointly by Lebenshilfe, the major German organization for parents of intellectually disabled infants, and the Bishop Bekkers Institute, a Dutch organization in the same field. The symposium was to be held in Marburg, a German university town, under the auspices of the International League of Societies for Persons with Mental Handicap, and the International Association for the Scientific Study of Mental Deficiency. The program looked impressive; after an opening speech from the German minister of family affairs, the conference was to be addressed by leading geneticists, bioethicists, theologians, and health-care lawyers from the United States, Canada, the Netherlands, England, France, and, of course, Germany. I accepted the invitation; and since I was going to be in Germany anyway, I also accepted an invitation from Professor Christoph

Anstötz, professor of special education at the University of Dortmund, to give a lecture a few days later on the subject "Do Severely Disabled Newborn Infants Have a Right to Life?"

My intention in these lectures was to defend a view for which I have argued in several previously published works: that the parents of severely disabled newborn infants should be able to decide, together with their physician, whether their infant should live or die. If the parents and their medical adviser are in agreement that the infant's life will be so miserable or so devoid of minimal satisfactions that it would be inhumane or futile to prolong life, then they should be allowed to ensure that death comes about speedily and without suffering. Such a decision might reasonably be reached, if, for instance, an infant was born with anencephaly (the term means "no brain" and infants with this condition have no prospect of ever gaining consciousness); or with a major chromosomal disorder such as trisomy 18, in which there are abnormalities of the nervous system, internal organs, and external features, and death always occurs within a few months, or at most two years; or in very severe forms of spina bifida where an exposed spinal cord leads to paralysis from the waist down, incontinence of bladder and bowel, a buildup of fluid on the brain, and, often, mental retardation. (Were these conditions to be detected in prenatal examinations, many mothers would choose to have an abortion, and their decision would be widely seen as understandable.)

Parents may not always be able to make an unbiased decision concerning the future of their infant, and their decisions may not be defensible. In some cases—Down syndrome, perhaps—the outlook for the child might be for a life without suffering, but the child would need much more care and attention, over a longer period, than a normal child would require. Some couples, feeling that they were not in a position to provide the care required, or that it would be harmful for their already existing family for them to try to do so, might oppose sustaining the infant's life. There may, however, be other couples willing to give the child an adequate home; or the community may be in a position to take over the responsibility of providing medical care and ensuring that the child has reasonably good conditions for living a satisfying life and developing his or her potential. In these circumstances, given that the child will not be living a life of unredeemed misery, and the parents will not be coerced into rearing that child, they can no longer insist upon having the major role in life-or-death decisions for their child.[10]

This position is, of course, at odds with the conventional doctrine of the sanctity of human life; but there are well-known difficulties in defending that doctrine in secular terms, without its traditional religious underpinnings. (Why, for example, if not because human beings are made in the image of God, should the boundary of sacrosanct life match the boundary of our species?) Among philosophers and bioethicists, the view that I was to defend is by no means extraordinary; if it has not quite reached the level of orthodoxy, it, or at least something akin to it, is widely held, and by some of the most respected scholars in the fields of both bioethics and applied ethics.[11]

JUST A DAY OR TWO before I was due to leave for Germany, my invitation to speak at the Marburg conference was abruptly withdrawn. The reason given was that, by agreeing to lecture at the University of Dortmund, I had allowed opponents of my views to argue that Lebenshilfe was providing the means for me to promote my views on euthanasia in Germany. The letter withdrawing the invitation drew a distinction between my discussing these views "behind closed doors with critical scientists who want to convince you that your attitude infringes human rights" and my promoting my position "in public." A postscript added that several organizations of handicapped persons were planning protest demonstrations in Marburg and Dortmund against me, and against Lebenshilfe for having invited me. (Although organizations for the disabled were prominent among the protesters, these groups were strongly supported and encouraged by various coalitions against genetic engineering and reproductive technology, and also by organizations on the left that had, apparently, nothing to do with the issue of euthanasia. The "Anti-Atom Bureau," for instance, joined the protests, presumably neither knowing nor caring about my opposition to uranium mining and nuclear power.)

The protests soon found their way into the popular press. *Der Spiegel*, which has a position in Germany not unlike that of *Time* and *Newsweek* in the United States, published a vehement attack on me written by Franz Christoph, the leader of the self-styled "Cripples Movement," a militant organization of disabled people.[12] The article was illustrated with photographs of the transportation of "euthanasia victims" in the Third Reich, and of Hitler's "Euthanasia Order." The article gave readers no idea at all of the ethical basis on which I advocated euthanasia, and it quoted spokespeople for groups of the disabled who appeared to believe that I questioned their right to life. I sent a brief reply in which I pointed out that I was advocating euthanasia not for anyone like themselves, but for severely disabled newborn infants, and that it was crucial to my defense of euthanasia that these infants would never have been capable of grasping that they are living beings with a past and a future. Hence my views cannot be a threat to anyone who is capable of wanting to go on living, or even of understanding that his or her life might be threatened. After a long delay, I received a letter from *Der Spiegel* telling me that, for reasons of space, they had been unable to publish my reply. Shortly afterward, however, *Der Spiegel* found space for a further highly critical account of my position on euthanasia, together with an interview, spread over four pages, with one of my leading opponents—and again, the same photograph of the Nazi transport vehicles.[13]

If Lebenshilfe had thought that they could pacify their critics by withdrawing my invitation to speak at Marburg, they had underestimated the storm that had broken loose. The protesters continued their opposition to what they were now calling the "Euthanasia Congress." Shortly before the symposium was due to open, Lebenshilfe and the Bishop Bekkers Institute canceled the entire event. Soon afterward the Faculty of Special Education at the University of Dortmund decided not to proceed with my scheduled lecture there.

THIS WAS NOT QUITE the end of my experiences in Germany that summer. Dr. Georg Meggle, professor of philosophy at the University of Saarbrücken, invited me to lecture at his university in order to show that it was possible to discuss the ethics of euthanasia rationally in Germany. I hoped to use this opportunity to say that, while I understood and strongly supported every effort to prevent the resurgence of Nazi ideas, my own views about euthanasia had nothing whatsoever to do with what the Nazis did. In contrast to the Nazi ideology that the state should decide who was worthy of life, my view was designed to reduce the power of the state and allow parents to make crucial life-and-death decisions, both for themselves and, in consultation with their doctors, for their newborn infants. Those who argued that it is always wrong to decide that a human life is not worth living would, to be consistent, have to say that we should use all the techniques of modern medical care in order to extend to the greatest possible extent the life of every infant, no matter how hopeless the infant's prospects might be and no matter how painful his or her existence. This was surely too cruel for any humane person to support.

Making this obvious point proved more difficult than I had expected. When I rose to speak in Saarbrücken, I was greeted by a chorus of whistles and shouts from a minority of the audience determined to prevent me from speaking. Professor Meggle offered the protesters the opportunity to state why they thought I should not speak. This showed how completely they had misunderstood my position. Many obviously believed that I was politically on the far right. Another suggested that I lacked the experience with Nazism that Germans had had; he and others in the audience were taken aback when I told them that I was the child of Austrian-Jewish refugees, and that three of my grandparents had died in Nazi concentration camps. Some seemed to think that I was opposed to all measures that would advance the position of the disabled in society, whereas in fact, while I hold that some lives are so severely blighted from the beginning that they are better not continued, I also believe that once a life has been allowed to develop, then in every case everything should be done to make that life as satisfying and rich as possible. This should include the best possible education, adjusted to the needs of the child, to bring out to the maximum the particular abilities of the disabled person.

Another chance comment revealed a still deeper ignorance about my position. One protester quoted from a passage in which I compare the capacities of intellectually disabled humans and nonhuman animals. The way in which he left the quotation hanging, as if it were in itself enough to condemn me, made me realize that he thought that I was urging that we should treat disabled humans in the way we now treat nonhuman animals. He had no idea that my views about how we should treat animals are utterly different from those conventionally accepted in Western society. When I replied that, for me, to compare a human being to a nonhuman animal was not to say that the human being should be treated with less consideration, but that the animal should be treated with more, this person asked why I did not use my talents to write about the morality of our treatment of animals, rather than about euthanasia. Naturally I replied that I had done that, and that it was, indeed, precisely for my views about the suffering of animals raised on commercial farms, and used in medical and psychological research, and the need for animal liberation that I was best known in English-speaking countries; but I could see that a large part of the audience simply did not believe that I could be known anywhere as anything other than an advocate of euthanasia.[14]

Allowing these misconceptions to be stated did, at least, provide an opportunity for reply. Someone else came to the platform and said that he agreed that it was not necessary to use intensive care medicine to prolong every life, but allowing an infant to die was different from taking active steps to end the infant's life. That led to further discussion, and so in the end we had a long and not entirely fruitless debate. Some of that audience, at least, went away better informed than they had been when they arrived.[15]

THE EVENTS OF THE SUMMER of 1989 have had continuing repercussions on German intellectual life. On the positive side, those who had sought to stifle the controversy over euthanasia soon found that, as so often happens, the attempt to suppress ideas only ensures that the ideas gain a wider audience. Germany's leading liberal weekly newspaper, *Die Zeit*, published two articles that gave a fair account of the arguments for euthanasia, and also discussed the taboo that had prevented open discussion of the topic in Germany. For this courageous piece of journalism, *Die Zeit* also became the target of protests, with Franz Christoph, the leader of the "Cripples Movement," chaining his wheelchair to the door of the newspaper's editorial offices. The editors of *Die Zeit* then invited Christoph to take part in a tape-recorded discussion with the editors of the newspaper and one or two others about whether the paper was right to discuss the topic of euthanasia. Christoph accepted, and the transcript was published in a further extensive article. Predictably, as in Saarbrücken, what began as a conversation about whether or not euthanasia should be discussed very soon turned into a debate on euthanasia itself.

From this point the euthanasia debate was picked up by both German and Austrian television. The outcome was that instead of a few hundred people hearing my views at lectures in Marburg and Dortmund, several million read about them or listened to them on television. The *Deutsche Ärzteblatt*—the major German medical journal—published an article by Helga Kuhse entitled "Why the Discussion of Euthanasia Is Unavoidable in Germany Too," which led to an extensive debate in subsequent issues.[16] In philosophical circles the discussion of applied ethics in general, and euthanasia in particular, is much livelier now than it was before 1989—as is indicated by the special issue of *Analyse & Kritik* to which I have already referred. In journals of special education, as well, ethical issues are now being discussed far more frequently than they were two years ago.

The protest also revived the flagging sales of the German edition of *Practical Ethics*. The book sold more copies in the year after June 1989 than it had in all the five years it had previously been available in Germany. Now everyone involved in the debate in Germany seems to be rushing to publish a book on euthanasia. With the exception of two books by Anstötz and Leist, which contain genuine ethical arguments, those published so far are of some interest for those wishing to study the thinking of Germans opposed to free speech, but not for any other reason.[17] For the most part each of the books appears to have been written to a formula that goes something like this:

1. Quote a few passages from *Practical Ethics* selected so as to distort the book's meaning.
2. Express horror that anyone can say such things.
3. Make a sneering gibe at the idea that this could pass for philosophy.
4. Draw a parallel between what has been quoted and what the Nazis thought or did.

But it is also essential to observe one negative aspect of the formula:

5. Avoid discussing any of the following dangerous questions: Is human life to be preserved to the maximum extent possible? If not, in cases in which the patient cannot and never has been able to express a preference, how are decisions to discontinue treatment to be made, without an evaluation of the patient's quality of life? What is the moral significance of the distinction between bringing about a patient's death by withdrawing treatment necessary to prolong life and bringing it about by active intervention? Why is advocacy of euthanasia for severely disabled infants so much worse than advocacy of abortion on request that the same people can oppose the right even to discuss the former, while themselves advocating the latter?

The irony about the recent publications, of course, is that even those who are highly critical of my own position do, by publishing their books and articles, foster a climate of debate about the topic. Even Franz Christoph, despite chaining his wheelchair to the door of *Die Zeit* because it published reports of my views on euthanasia, has now published his own book on the topic. At the outset he protests vigorously that his book is not a contribution *to* the debate about euthanasia, but a book *against* this debate; it is self-evident, though, that one cannot publish a book on whether or not to have a debate on euthanasia without stimulating thought among one's readers and reviewers about the issue of euthanasia itself.[18]

THE NEGATIVE ASPECTS of these events are, unfortunately, probably more weighty. Most threatening of all are the incidents described at the beginning of this essay, and the atmosphere of repression and intimidation that they have evoked. Anyone who offers a course based on *Practical Ethics* in Germany now risks the same protests and personal attacks that Professor Kliemt faced in Duisburg. One philosopher in Berlin told me recently that it is not possible to offer a course in applied ethics in that city—whether or not it makes reference to my book—because such a course would be bound to be disrupted.

A sinister aspect of this atmosphere is a kind of self-censorship among German publishers. It has proved extraordinarily difficult to find a publisher to undertake a German edition of *Should the Baby Live?*—the updated and more comprehensive account of my views (and those of my coauthor Helga Kuhse) on the treatment of severely disabled newborn infants. In view of the current controversy, there seems no doubt that a German edition of the book would have good commercial prospects. Yet one after another, German publishers have declined to publish it, even after it had been recommended by editors whose advice they normally accept without hesitation.[19]

For those interested in studying or teaching bioethics or applied ethics in Germany, the consequences are much more serious still. Because he had invited me to lecture at the University of Dortmund, Professor Christoph Anstötz became the target of a hostile campaign aimed at having him dismissed from his teaching duties. Petitions were circulated and letters written to the minister of science and research for the state of Nordrhein-Westfalen, in which Dortmund is situated. These letters were signed by both teachers and students in special education. Although Professor Anstötz has a tenured position from which it would scarcely be possible for him to be dismissed, the government took the complaints seriously enough to ask him to explain why he had invited me, and what implications he drew from my ethical position for his work in special education.

Throughout this campaign, the rector of the University of Dortmund and his office remained silent. The highest officers of the university took no action to indicate their concern that threats of protest had forced an academic lecture to be canceled; nor did they come to the defense of one of their professors when he was under attack for inviting a colleague to give a lecture on the campus of the university. That was typical of the reaction of German professors. There was no strong reaction among them on behalf of academic freedom. With a handful of exceptions, Anstötz's colleagues in special education either joined the campaign against him or remained silent. A number of philosophers signed declarations of support for the principle of free debate, and one of these was published in the Berlin newspaper *taz*.[20] At Professor Meggle's instigation, 180 members of the German Philosophical Association signed a similar declaration, but the association has since failed to publish the list of the signers, despite giving an undertaking to do so.

All this does not augur well for the future of rational discussion of controversial new ethical issues in Germany and Austria. Outside the German-speaking nations, study and discussion of bioethics are expanding rapidly, in response to the recognition of the need for ethical consideration of the many new issues raised by developments in medicine and the biological sciences. Other fields of applied ethics, such as the status of animals, questions of global justice and resource distribution, environmental ethics, and business ethics, are also getting much attention. In Germany and Austria, however, it now takes real courage to do work in applied ethics, and even more courage to publish something that is likely to come under the hostile scrutiny of those who want to stop debate. Academics who do not have a permanent university position must fear not merely personal attack but also the diminished opportunity to pursue an academic career. The events in Hamburg cast a cloud over the prospect that university posts will open up in these fields. If there are no posts to be obtained, graduate students will avoid working on questions of applied ethics, for there is no sense in studying matters that offer no prospect of employment. There is even a danger that in order to avoid controversy, analytic philosophy as a whole will suffer a setback. At the present time, a large number of new university positions are being created in the universities of the former German Democratic Republic. Philosophers interested in analytic philosophy are concerned that these positions may all go to philosophers working on less sensitive subjects, for example, to those who concentrate on historical studies, or to followers of Habermas who have generally kept quiet about these sensitive ethical issues and about the obstacles to debating them in Germany today.

GERMANS, OF COURSE, ARE STILL STRUGGLING to deal with their past, and the German past is one which comes close to defying rational understanding. There is, however, a peculiar tone of fanaticism about some sections of the German debate over euthanasia that goes beyond normal opposition to Nazism, and instead begins to seem like the very mentality that made Nazism possible. To see this attitude at work, let us look not at euthanasia but at an issue that is, for the Germans, closely related to it and just as firmly taboo: the issue of eugenics. Because the Nazis practiced eugenics, anything in any way related to genetic engineering in Germany is now smeared with Nazi associations. This attack embraces the rejection of prenatal diagnosis, when followed by selective abortion of fetuses with Down syndrome, spina bifida, or other defects, and even leads to criticism of genetic counseling designed to avoid the conception of children with genetic defects. It has also led to the German parliament's unanimously passing a law that prohibits all nontherapeutic experimentation on the human embryo. The British parliament, by contrast, recently passed by substantial majorities in both chambers a law that allows nontherapeutic embryo experimentation up to fourteen days after fertilization.

To understand how bizarre this situation is, readers in English-speaking countries must remind themselves that this opposition comes not, as it would in our countries, from right-wing conservative and religious groups, but from the left. Since women's organizations are prominent among the opposition to anything that smacks of eugenics, and also are in the forefront of the movement to defend the right to abortion, the issue of prenatal diagnosis gives rise to an obvious problem in German feminist circles. The accepted solution seems to be that a woman should have the right to an abortion, but not to an abortion based on accurate information about the future life prospects of the fetus she is carrying.[21]

The rationale for this view is, at least, consistent with the rationale for opposition to euthanasia: it is the idea that no one should ever judge one life to be less worth living than another. To accept prenatal diagnosis and selective abortion, or even to select genetic counseling aimed at avoiding the conception of infants with extreme genetic abnormalities, is seen as judging that some lives are less worth living than others. To this the more militant groups of disabled people take offense; it suggests, they maintain, that they should not have been allowed to come into existence, and thus denies their right to life.

This is, of course, a fallacy. It is one thing to hold that we may justifiably take steps to ensure that the children we bring into the world do not face appalling obstacles to living a minimally decent life, and a quite different thing to deny to a living person who wants to go on living the right to do just that. If the suggestion, on the other hand, is that whenever we seek to avoid having severely disabled children, we are improperly judging one kind of life to be worse than another, we can reply that such judgments are both necessary and proper. To argue otherwise would seem to suggest that if we break a leg, we should not get it mended, because in doing so we judge the lives of those with crippled legs to be less worth living than our own.[22] For people to believe such a fallacious argument is bad enough; what is really frightening, however, is that people believe in it with such fanaticism that they are prepared to use force to suppress any attempt to discuss it.

If this is the case with attempts to discuss practices like genetic counseling and prenatal diagnosis, which are today very widely accepted in most developed countries, it is easy to imagine that the shadow of Nazism prevents any rational discussion of anything that relates to euthanasia. It avails little to point out that what the Nazis called "euthanasia" had nothing to do with compassion or concern for those who were killed but was simply the murder of people considered unworthy of living from the racist viewpoint of the German *Volk*. Such distinctions are altogether too subtle for those who are convinced that they alone know what will prevent a revival of Nazi-like barbarism.

CAN ANYTHING BE DONE? In May this year, in Zurich, I had one of the most unpleasant experiences yet in this unhappy story; but it gave, at the same time, a glimmer of hope that there may be a remedy.

I was invited by the Zoological Institute of the University of Zurich to give a lecture on animal rights. On the following day, the philosophy department had organized a colloquium for twenty-five invited philosophers, theologians, special educationalists, zoologists, and other academics to discuss the implications for both humans and animals of an ethic that would reject the view that the boundary of our species marks a moral boundary of great intrinsic significance, and holds that nonhuman animals have no rights.

The lecture on animal rights did not take place. Before it began, a group of disabled people in wheelchairs, who had been admitted to the flat area at the front of the lecture theater, staged a brief protest in which they said that, while it was all the same to them whether or not I lectured on the topic of animal rights, they objected to the fact that the University of Zurich had invited such a notorious advocate of euthanasia to discuss ethical issues that also concerned the disabled. At the end of this protest, when I rose to speak, a section of the audience—perhaps a quarter or a third—began to chant: "Singer *raus!* Singer *raus!*" As I heard this chanted, in German, by people so lacking in respect for the tradition of reasoned debate that they were unwilling even to allow me to make a response to what had just been said about me, I had an overwhelming feeling that this was what it must have been like to attempt to reason against the rising tide of Nazism in the declining days of the Weimar Republic. The difference was that the chant would have been, not "Singer *raus,*" but "*Juden raus.*" An overhead projector was still functioning, and I began to write on it, to point out this parallel that I was feeling so strongly. At that point one of the protesters came up behind me and tore my glasses from my face, throwing them on the floor and breaking them.

My host wisely decided to abandon the lecture; there was nothing else that could be done. But from this distressing affair came one good sign; it was clear that the disabled people who had made the initial protest were distressed with what had happened afterward. Several said that they had not intended that the lecture should be disrupted; they had, in fact, prepared questions to ask during the discussion period that would have followed the lecture. Even while the chanting was going on, some attempted to begin a discussion with me; at which point some of the able-bodied demonstrators (presumably well aware of the way in which in Saarbrücken a discussion had broken through the initial hostility toward me) urgently remonstrated with them not to talk to me. The disabled, however, clearly had no power to do anything about the chanting.

As already noted, my views in no way threaten anyone who is, or ever has been, even minimally aware of the fact that he or she has a possible future life that could be threatened. But there are some who have a political interest in preventing this elementary fact from becoming known. These people are now playing on the anxieties of the disabled in order to use them as a political front for different purposes. In Zurich, for instance, prominent among the nondisabled people chanting "Singer *raus*" were the *Autonomen,* or "Autonomists," a group that affects an anarchist style but disdains any interest in anarchist theory. For these nondisabled political groups, preventing Singer from speaking, no matter what the topic, has become an end in itself, a way of rallying the faithful and striking at the entire system in which rational debate takes place. Disabled people have nothing to gain, and much to lose, by allowing themselves to be used by such nihilistic groups. If they can be brought to see that their interests are better served by an open discussion with those whose views they oppose, it may be possible to begin a process in which both bioethicists and the disabled address the proper concerns of the other side and move to a dialogue that is constructive rather than destructive.

SUCH A DIALOGUE would be only a beginning. To heal the damage done to bioethics and applied ethics in Germany will take much longer. There is a real danger that the atmosphere of intimidation and intolerance which has spread from the issue of euthanasia to all of bioethics, and, with the events in Hamburg, to applied ethics in general, will continue to broaden. It is essential that the minority that is actively opposing the free discussion of academic ideas be isolated. Here too, what happened in Zurich may serve as an example for other German-speaking countries to follow. In sharp contrast to the silence of the rector of the University of Dortmund, or the fatuous claim that "We didn't know at all who that was" of the dean of medicine at the University of Vienna, Professor H. H. Schmid, rector of the University of Zurich, issued a statement expressing the university's "outrage over this grave violation of academic freedom of speech."[23] The professors of the Zoological Institute and the dean of the Faculty of Science have also unequivocally condemned the disruption, and the major German-language newspapers in Zurich gave objective coverage to the events and to my views.[24]

Meanwhile Germans and Austrians, both in academic life and in the press, have shown themselves sadly lacking in the commitment exemplified by the celebrated utterance attributed to Voltaire: "I disapprove of what you say, but I will defend to the death your right to say it." No one has, as yet, been asked to risk death in order to defend my right to discuss euthanasia in Germany, but it is important that many more should be prepared to risk a little hostility from the minority that is trying to silence a debate on central ethical questions.

An Interview

INTERVIEWS WITH JOURNALISTS are often dismissed by academics, because the pressure to simplify one's ideas is so great—both because time is short, and in order to make complex ideas comprehensible to an audience not used to taking in long chains of abstract reasoning. Interviews do need to be treated with care, because the interviewee usually has no control over the final editing, which can remove important qualifications or imply that a comment relates to something altogether different from the context in which it was made. Nevertheless the need to respond to the questions and challenges of interviewers provides an opportunity to clarify misunderstandings and meet objections. In that spirit, I have included below a composite "interview," which is an amalgamation of two separate interviews given in the aftermath of my appointment to Princeton. To an interview by Bob Abernethy for the Public Broadcasting Service program "Religion and Ethics Newsweekly," shown on WNET-TV on September 10, 1999, I have added some further questions asked by Nell Boyce, for an interview that appeared in *New Scientist*, January 8, 2000. Where I found my responses to their questions less felicitous than I would have wished, I have taken the opportunity to clarify or amplify what I said.

BOB ABERNETHY: Let's start with some of your basic ideas. You say that a human being, a person, doesn't necessarily have value because of some intrinsic quality of just being a person; but what's important are certain qualities. What are those qualities?

Peter Singer: First, it is important to say that in my view it is a *human being*, not a person, who doesn't have value simply in virtue of belonging to the species *Homo sapiens*. Species membership alone isn't enough. The qualities that I think are important are, first, a capacity to experience something—that is, a capacity to feel pain, or to have any kind of feelings. That's really basic. But then that's something we share with a huge range of nonhuman animals. In addition, when it comes to a question of taking life, or allowing life to end, I would say it matters whether a being is the kind of being who can see that he or she actually has a life—that is, can see that he or she is the same being who exists now, who existed in the past, and who will exist in the future.

I use the term "person" to refer to a being with that kind of self-awareness—in the words of the philosopher James Rachels, a being who can live a *biographical* life and not merely a *biological* life. A person has a lot more to lose when his or her life is ended than a being that is conscious, and can feel pain, but nevertheless is conscious of its existence only moment by moment, experiencing only one moment of consciousness and then the next, without understanding the connection between them.

BA: Some of your critics have accused you—on just that point—of abandoning the entire Judeo-Christian tradition regarding the value of human life. What do you say to that?

PS: I accept the accusation. I think that the Judeo-Christian tradition has an unjustifiable bias in favor of human beings *qua* human beings; to that extent it needs far-reaching revision. If you look at the book of Genesis, you see there the idea that humans are special, that God created humans in his own image and gave them dominion over the other animals. Since Darwin, at least, we've known that that's factually false, and now we've got to draw the moral implications of understanding that it's factually false.

BA: Okay, the idea that many of us were raised with and cling to is that each human being is a creature of God, has intrinsic worth because of that, and that there is a sanctity, therefore, to human life. How do you deal with that?

PS: I don't believe in the existence of God, so I also reject the idea that each human being is a creature of God. It's as simple as that. Now, if you have a different view, coming from a religious belief, obviously, you're entitled to live your own life in accordance with your religious views, as long as you don't interfere with others who do not share those views. What I would like to see is a society that, in its laws and public ethics, was not dominated by any specifically religious doctrines. Clearly, not everyone in this society or any other society that I've lived in does believe in God.

BA: So, how do you arrive at what is moral? What is your basic, underlying principle?

PS: We have to use our own thought and reflection to try and see what we would ourselves object to if it were done to us. To that extent, I could say that I am part of a religious tradition in that this is very like the golden rule. But, of course, the golden rule is not exclusive to the Judeo-Christian tradition. You find it in other religious traditions, too. I think it's something that thinking humans can come to independently of religion—this idea of not wanting to do to others what you wouldn't like to have done to yourself. Perhaps, as R. M. Hare, my teacher at Oxford University, argued, it is something that can be derived from the very concept of morality. In any case, I see it as a basis from which you can develop a moral outlook, and that's what I've tried to do.

BA: But you also say that determining whether an act is moral depends on its consequences.

PS: I think that the idea of determining right and wrong by looking at the consequences is something that can flow from the idea of the golden rule, although certainly not all thinkers have taken it that way. But if you say, "If I were in that position, I wouldn't want that done to me," you are, in fact, looking at the consequences of the act. You're not looking at whether or not it conforms with a rule.

BA: Just to be clear about this, yours is essentially a utilitarian position, that whether something is moral depends on what the likely outcome is. Right?

PS: Right. I don't think you can decide whether an action is right or wrong without looking at what its effects are, what does it do, what impact does it have on people, animals, or the planet.

BA: Now, where does all this lead? You care very deeply and have written a great deal about relieving suffering. Talk about that a bit. Not only individual suffering, but in terms of the whole globe.

PS: I think that if we follow that idea of "doing unto others," then, even though people have different sorts of preferences and different wants, one thing is pretty general: people do not want to suffer. They do not want extreme physical pain; they do not want emotional deprivation and suffering. That's something that we share with nonhuman animals, broadly. And it's also something that we in prosperous countries like the United States share with people in the developing world. Now, there are various ways in which we could quite easily reduce the amount of suffering that there is in the universe. One way relates to animals. We could stop doing a lot of things to animals which cause them to suffer, and which we don't need to do. That would reduce their suffering. Another way relates to people in the world's poorest countries. We could give some of the superfluous wealth that we fortunate people living in the affluent nations have, and use it to relieve the terrible suffering of people who are so poor that they are living on the edge of malnutrition and dying from easily preventable diseases. A third way of reducing suffering would be to assist people who are dying from diseases like cancer, who are in pain and distress, and who say, "Look, I've had enough. I don't want to go on." We could enable them to act upon their own decision about when they've had enough. The views of mine that have been most controversial all stem from this idea that we should reduce the amount of suffering in the world, if we can do so.

BA: Talk a little more about how it would be appropriate to help the old and the sick. When and why would it be all right to kill an old person—a sick, old person?

PS: It would be all right to kill a sick, old person when that person has asked to be killed, when that request has been made clearly and persistently; and when we are convinced that the person is in a sound, rational frame of mind and is making that decision for good reason—such as the fact that he or she has terminal cancer.

BA: I suppose the idea that has provoked the most controversy is your belief that it is all right under some circumstances to take the life of a newborn child. Would you say just as clearly and precisely as you possibly can what your position is on that?

PS: There are some disabled infants born with conditions so severe that doctors don't really try to keep them alive. They allow them to die, essentially through benign neglect. But that can be a very slow process. In my view, if that decision is justified—and I think it can be—then we should not simply allow the child to die from neglect. With the consent and support of the parents, advised by their doctors—and only then—I think it would be justifiable to help that infant to die. It would be justifiable to take active steps to end that infant's life swiftly and more humanely than by allowing death to come through dehydration, starvation, or an untreated infection.

BA: You said in a quotation that has showed up in many places: "Killing a defective infant is not morally equivalent to killing a person. Sometimes it is not wrong at all." Do you want to expand on that?

PS: As I said earlier, I use the term "person" to refer to a being who is capable of anticipating the future, of having wants and desires for the future. If that person is killed against his or her will, these desires are cut off, thwarted. For this reason, among others, I think that it is generally a greater wrong to kill a person than it is to kill a being that has no sense of existing over time. Perhaps, for example, a chicken has no sense of existing over time. And that, I think, is one reason why it's normally worse to kill a human being than to kill a chicken. But newborn human babies have no sense of their own existence over time. So killing a newborn baby—whether able-bodied or not—is never equivalent to killing a person, that is, a being who wants to go on living. It's different. That doesn't mean that it is not almost always a terrible thing to do. It is, but that is because most infants are loved and cherished by their parents, and to kill an infant is usually to do a great wrong to its parents.

BA: Are you worried about a slippery slope there—that if it is permitted to kill disabled newborns, this might be extended in some way to others?

PS: We have to be aware of the slippery-slope problem, and we have to think carefully about it. But it's not as if we're now on flat ground, and accepting my view would be the first step onto a slippery slope. We're already on that slope. For example, by allowing the termination of pregnancy, we have taken a step that violates the traditional view of the sanctity of human life. Many people will say, "Well, that's why abortion is wrong." But what about the reclassification of people as dead when the brain has irreversibly ceased to function? That has been accepted in this country for more than twenty years without any serious opposition, but it's also a step down the slippery slope. Once you can reclassify a person as dead because the brain has ceased to function, even though the body is warm and the heart is beating, you could go further. With modern medical techniques, there's no way in which we can just say, "You must never end the life of another human being." We're already doing it all the time, either by abortion, or by withdrawing or withholding life-supporting treatment, or by categorizing the being as brain-dead. So the question is not, "Can we stay off the slippery slope?" but, rather, "How can we best negotiate the slippery slope so as not to slide anywhere we don't want to go?"

NELL BOYCE: Your grandfather and other members of your family died in the Holocaust. When people call you a Nazi, do you ignore it, or do you worry that some of your ideas may have unforeseen consequences?

PS: I find the throwing around of the "Nazi" accusation offensive. It sadly trivializes the enormity of the Nazi crimes. And it's just absurd, because I come from a totally different political direction. I'm a social democrat, utterly opposed to racist policies and a totalitarian state. But do I reflect on the idea that some of the things I've said could lead in a direction that would not be a force for good? Yes, certainly I think about, say, the risk that my views may make society less supportive of people with disabilities. But I don't think the right answer is to say: "Well, we must not challenge the traditional ethic of the sanctity of human life," even though we can see that it's actually founded on fictions or outmoded views of the world. I think that if you try to cover up the cracks in the ethic, you're more likely to get a major crash in the long run.

NB: What kind of crash?

PS: The traditional ethic of the sanctity of life is being eroded on all sides by practices related to medical technology, such as advanced life-support systems. In the future, we may end up paying lip service to the ethic. Eventually people could simply abandon it, but they won't have anything else to put in its place. The result could be complete confusion about what might make killing wrong in any circumstances. The traditional ethic is not sustainable. There are other ways of looking at the wrongness of killing, which show why killing is wrong in, for example, the case of any self-aware being who wants to go on living. A principle like that, widely understood, is more likely to be successful in preventing things like the Holocaust than sticking to an ethic that really makes sense only within the context of a Judeo-Christian worldview.

NB: Do you feel that any of your ideas have been misrepresented?

PS: Yes, especially my views regarding euthanasia for disabled infants. The misrepresentation is of various kinds, but it usually comes from taking a sentence or two from *Practical Ethics*, which was written as a textbook for university use, and suggesting that this is my view or that I think it should immediately be put into practice as public policy. Very often what I am doing is following the implications of various ethical views and getting students to think about whether they accept these implications.

NB: But haven't you argued that parents should be allowed to kill a disabled infant or even one with a treatable disease such as hemophilia if it allows them to have a child with a greater chance of happiness?

PS: Hemophilia is one of the misrepresented examples I was referring to. The quotation everyone uses is plucked from a section in *Practical Ethics* in which I was developing the implications of a particular view of utilitarianism to get people to think about the differences between that view and an alternative view. I wasn't suggesting as a matter of public policy that parents should be allowed to kill infants with hemophilia. In our society today, that would be wrong. Hemophilia is not the disastrous condition that it once was, and it is hard to imagine that parents would really wish to kill a child who had it. But if, for some strange reason, they did think that they could not cope with their child, it would not be difficult to find a childless couple who would be delighted to adopt such a child. Killing it should not be an option.

BA: Another group of people concerned by some of your ideas have been handicapped people who, perhaps totally mistakenly, see in some of the things that you have written about a possible threat to people like them.

PS: That's a misunderstanding of my views, and a particularly unfortunate one, because it has caused distress to some people with disabilities. I have written that every disabled person should be supported in trying to live the best possible life that he or she can, as long as he or she wants to do so—as with all of the rest of us. And I regret, in fact, that the facilities for disabled people are not better than they are, that we don't give them more support. It's certainly nothing against people with disabilities that motivates my position. It is, rather, a desire to avoid suffering that is unnecessary, that can be avoided, right at the outset of life, at the stage of a newborn infant. That is obviously not a threat to any person with a disability who is capable of understanding anything about my position.

NB: Has your experience with your mother, who is profoundly disabled with Alzheimer's disease, influenced your views about creatures with a limited capacity for self-awareness?

PS: I couldn't say that it's totally unrelated, but I don't think it's had an impact. My mother is not suffering from her condition, because she lacks the self-awareness that would lead her to suffer from it. So it's not like the cases of euthanasia that I've written about.

NB: Years ago, an infant was born in Bloomington, Indiana, with both Down syndrome and a defective digestive tract which the parents decided not to surgically correct, so the child died. Are similar decisions being made now to kill infants through neglect?

PS: Probably there is now a more widespread acceptance that people with Down syndrome can have good lives and that parents should be encouraged to allow surgery even if they are going to give the child up for adoption. But certainly decisions to withhold life-prolonging medical treatment still get made in cases where the surgery would be complicated and the outcome uncertain, or where the underlying conditions are more severely disabling. In the United States, there is on the whole a much more aggressive treatment of these cases than there would be in Britain, Australia, or many other countries.

NB: Are you satisfied with the situation?

PS: No, I'm not satisfied. I don't think it's a good situation because the law doesn't really clarify what doctors are entitled to do. There are certain things going on which are legally dubious but which may be right. The fact that doctors have to do the right thing in a secretive way is not a good thing, because it means that doctors and parents can't be totally open with each other.

BA: Let's extend this now to animals. Just sketch, if you would, the moral case for what you've called "animal liberation."

PS: The case for animal liberation is very simple. It's that animals can feel and have interests. There is no reason why we should give less consideration to their interests than we give to similar interests of members of our own species. The fact that animals are not members of our species is, in itself, no more morally relevant than the fact that a human being is not a member of my race or not a member of my sex.

BA: And, again, back to those qualities we spoke about earlier?

PS: The key quality that animals share with us is the capacity to feel pain and the capacity to suffer. And, therefore, they have an interest in not suffering. Some of them may share other qualities. Chimpanzees and orangutans may be "persons" in the sense that I mentioned. They may be capable of seeing themselves as existing over time. A lot of animals—including some animals we eat, like chickens and fish—may not be persons. I'm not sure about whether cows and pigs qualify as persons—I would prefer to give them the benefit of the doubt. But what's clear is that they can all suffer. And when we raise them for food, we ignore their capacities for suffering. We use them just as things, and we frustrate their most important needs in order to satisfy some quite minor needs of our own.

BA: And this is what you've called "speciesism?" Tell me about "speciesism."

PS: Speciesism is a term that I've used to make a parallel between racism and sexism on the one hand, and our attitude to animals on the other. The attitude of white racists to Africans was: "You're not a member of my race. Therefore, it's okay for me to capture you, to enslave you, to use you as a living tool to work my plantations." When you think about what we do to animals, it's quite similar. We say, "You're not a member of my species. Therefore, it's okay for me to capture you, to breed you, to make you into a thing, to use you as a tool for producing food or eggs or milk—or to use you as an experimental tool in the laboratory." The fact that a being is not a member of our species does not, in itself, justify doing any of those things to it.

BA: You referred to medical experiments, but that's a particularly important part of it, isn't it?

PS: No, not really. I think the use of animals for food is a greater moral wrong than the use of our animals in experiments, because it's clearly less necessary, and it involves a much larger number of animals.

BA: But on the question particularly of the medical experimentation, when you get into consideration of the consequences, one might be a cure for a disease that would relieve the suffering of a great many persons. Right?

PS: It might be. I have never said that I think all animal experimentation should stop immediately. If you can show that that is the only way of achieving a goal like curing a major disease, then I would say we should seek alternative ways of getting to the same goal, but in the meanwhile, I would not campaign to stop those particular experiments. But what I do say is that, in general, animals have been used just as things, just as tools. They've been cheap. No one has cared about their interests. It's just been a matter of, "Let's order up another 200 mice or rats or guinea pigs for next Monday morning and try this out on them." So, that's where the speciesism comes in—the fact that we're prepared to do this to animals. We condemn doing that to humans, irrespective of their mental level.

NB: You've said that research on a chimp can be justified only when the experiment is so important that the use of a brain-damaged human would also be justifiable. So would it be OK to use brain-damaged humans?

PS: You would have to look at it on a case-by-case basis. I wouldn't absolutely rule it out. The point of what I said is that we are incredibly more protective of human beings than we are of nonhuman animals. Getting people to make that comparison makes them think about what kind of case for experimentation would be strong enough for us to say, "Yes, we really are prepared to do that experiment on a brain-damaged human." If the case is not strong enough to justify that conclusion, then how can it be strong enough to justify doing the research on a chimpanzee who is at a higher mental level than the human we have just said may not be experimented upon?

NB: As a pioneer of the modern animals rights movement, how do you now feel about activists sending scientists razor blades through the mail?

PS: I think that it's a deplorable thing to do. It risks serious damage to the movement because the movement's strength is the fact that it takes a strong moral stand and that it has a really good moral case. By using these tactics, the risk is that the movement will be seen simply as crazy extremists trying to force their views on other people.

BA: You've been called a lot of things that I'm sure you've learned to deal with, but it still must sting. Somewhere I read that somebody had called you the most dangerous man in the world today, and somebody else had used the phrase "Professor Death." Why do you think what you're saying has provoked such passion?

PS: What I'm saying is controversial. As we've been discussing, it goes against an ethic that has been around for a long time. That ethic is in the process of change, but I have, perhaps, brought out a little more clearly and bluntly than most the way in which that ethic needs to continue to change. And I've refused to try and disguise what I'm saying behind a veil which says, for example, "We're not killing; we're merely letting die," or "We're not cutting beating hearts out of living human beings, because the brains of these humans have ceased to function and so they are dead." My refusal to draw that delicate veil of euphemism over things to make them more acceptable is a large part of the reason why I get all this opposition while other people, whose conclusions are largely in agreement with mine, do not.

BA: How does it make you feel?

PS: I really do not like some of the things that I've been called. I find them unfair and inflammatory. They're basically media sound bites. But in so far as it helps me to get my ideas across to a wider audience—and it certainly does that—I can see that it has its good side as well.

NB: What projects are you working on these days?

PS: I am completing something that is completely different, a book about my grandfather, who lived in Vienna from the late nineteenth century until the Holocaust. I find it fascinating to go back to that period and piece together the life of a man I never knew. I'm getting close to the end of it now. That, plus the move to Princeton, has made it a good time to stop and decide what I really want to be working on next. I'll probably do something relating to genetics. I'm also interested in some of the global issues, like justice and world hunger, climate change, and the ethics of a global free market.

Notes

MORAL EXPERTS

1. A. J. Ayer, "The Analysis of Moral Judgments" in *Philosophical Essays* (London: Macmillan, 1954).
2. C. D. Broad, *Ethics and the History of Philosophy* (London: Routledge & Kegan Paul, 1952).
3. Ryle, "On Forgetting the Difference between Right and Wrong" in *Essays in Moral Philosophy*, ed. A. Melden (Seattle: University of Washington Press, 1958).

ALL ANIMALS ARE EQUAL . . .

1. For Bentham's moral philosophy, see his *Introduction to the Principles of Morals and Legislation*, and for Sidgwick's see *The Methods of Ethics*, 1907 (the passage is quoted from the seventh edition; reprint, London: Macmillan, 1963), p. 382. As examples of leading contemporary moral philosophers who incorporate a requirement of equal consideration of interests, see R. M. Hare, *Freedom and Reason* (New York: Oxford University Press, 1963), and John Rawls, *A Theory of Justice* (Cambridge: Harvard University Press, Belknap, 1972). For a brief account of the essential agreement on this issue between these and other positions, see R. M. Hare, "Rules of War and Moral Reasoning," *Philosophy and Public Affairs* 1, 2 (1972).
2. Letter to Henry Gregoire, February 25, 1809.
3. Reminiscences by Francis D. Gage, from Susan B. Anthony, *The History of Woman Suffrage*, vol. 1; the passage is to be found in the extract in

Leslie Tanner, ed., *Voices from Women's Liberation* (New York: Signet, 1970).

4. I owe the term "speciesism" to Richard Ryder. It has become accepted in general use since the first edition of this book and now appears in *The Oxford English Dictionary*, 2nd ed. (Oxford: Clarendon Press, 1989).

5. *Introduction to the Principles of Morals and Legislation*, chap. 17.

6. See M. Levin, "Animal Rights Evaluated," *Humanist* 37: 14–15 (July/August 1977); M. A. Fox, "Animal Liberation: A Critique," *Ethics* 88: 134–138 (1978); C. Perry and G. E. Jones, "On Animal Rights," *International Journal of Applied Philosophy* 1: 39–57 (1982).

7. Lord Brain, "Presidential Address," in C. A. Keele and R. Smith, eds., *The Assessment of Pain in Men and Animals* (London: Universities Federation for Animal Welfare, 1962).

8. Ibid., p. 11.

9. Richard Serjeant, *The Spectrum of Pain* (London: Hart Davis, 1969), p. 72.

10. See the reports of the Committee on Cruelty to Wild Animals (Command Paper 8266, 1951), paragraphs 36–42; the Departmental Committee on Experiments on Animals (Command Paper 2641, 1965), paragraphs 179–182; and the Technical Committee to Enquire into the Welfare of Animals Kept under Intensive Livestock Husbandry Systems (Command Paper 2836, 1965), paragraphs 26–28 (London: Her Majesty's Stationery Office).

11. See Stephen Walker, *Animal Thoughts* (London: Routledge and Kegan Paul, 1983); Donald Griffin, *Animal Thinking* (Cambridge: Harvard University Press, 1984); and Marian Stamp Dawkins, *Animal Suffering: The Science of Animal Welfare* (London: Chapman and Hall, 1980).

12. See Eugene Linden, *Apes, Men and Language* (New York: Penguin, 1976); for popular accounts of some more recent work, see Erik Eckholm, "Pygmy Chimp Readily Learns Language Skill," *New York Times*, June 24, 1985; and "The Wisdom of Animals," *Newsweek*, May 23, 1988.

13. *In the Shadow of Man* (Boston: Houghton Mifflin, 1971), p. 225. Michael Peters makes a similar point in "Nature and Culture," in Stanley and Roslind Godlovitch and John Harris, eds., *Animals, Men and Morals* (New York: Taplinger, 1972). For examples of some of the inconsistencies in denials that creatures without language can feel pain, see Bernard Rollin, *The Unheeded Cry: Animal Consciousness, Animal Pain, and Science* (Oxford: Oxford University Press, 1989).

14. I am here putting aside religious views, for example the doctrine that all and only human beings have immortal souls or are made in the image of God. Historically these have been very important, and no doubt are partly responsible for the idea that human life has a special sanctity. Logically, however, these religious views are unsatisfactory, since they do not offer a reasoned explanation of why it should be that all humans and no nonhumans have immortal souls. This belief too, therefore, comes under suspicion as a form of speciesism. In any case, defenders of the "sanctity of life" view are generally reluctant to base

their position on purely religious doctrines, since these doctrines are no longer as widely accepted as they once were.

15. For a general discussion of these questions, see my *Practical Ethics* (Cambridge: Cambridge University Press, 1979), and for a more detailed discussion of the treatment of handicapped infants, see Helga Kuhse and Peter Singer, *Should the Baby Live?* (Oxford: Oxford University Press, 1985). See also the section of this volume, "Saving and Taking Human Life."

16. For a development of this theme, see my essay, "Life's Uncertain Voyage," in P. Pettit, R. Sylvan, and J. Norman, eds., *Metaphysics and Morality* (Oxford: Blackwell, 1987), pp. 154–172.

17. The preceding discussion, which has been changed only slightly since the first edition, has often been overlooked by critics of the Animal Liberation movement. It is a common tactic to seek to ridicule the Animal Liberation position by maintaining that, as an animal experimenter put it recently, "Some of these people believe that every insect, every mouse, has as much right to life as a human." (Dr. Irving Weissman, as quoted in Katherine Bishop, "From Shop to Lab to Farm, Animal Rights Battle Is Felt," *New York Times*, January 14, 1989.) It would be interesting to see Dr. Weissman name some prominent animal liberationists who hold this view. Certainly (assuming only that he was referring to the right to life of a human being with mental capacities very different from those of the insect and the mouse) the position described is not mine. I doubt that it is held by many—if any—in the animal liberation movement.

TOOLS FOR RESEARCH

1. Since this was written, most cosmetics companies have, after campaigns from animal rights organizations, ceased to test their products on animals.

2. Report of the Littlewood Committee, pp. 53, 166; quoted by Richard Ryder, "Experiments on Animals," in Stanley and Roslind Godlovitch and John Harris, eds., *Animals, Men, and Morals* (New York: Taplinger, 1972), p. 43.

3. *Journal of Abnormal and Social Psychology* 48 (2): 291 (April 1953).

4. *Journal of Abnormal Psychology* 73 (3): 256 (June 1968).

5. *Animal Learning and Behavior* 12: 332–338 (1984).

6. *Journal of Experimental Psychology: Animal Behavior and Processes* 12: 277–290 (1986).

7. *Psychological Reports* 57: 1027–1030 (1985).

8. *Progress in Neuro-Psychopharmacology and Biological Psychiatry* 8: 434–446 (1984).

9 H. Beecher, "Ethics and Clinical Research," *New England Journal of Medicine* 274: 1354–1360 (1966); D. Rothman, "Ethics and Human Experimentation: Henry Beecher Revisited," *New England Journal of Medicine* 317: 1195–1199 (1987).

10. T. McKeown, *The Role of Medicine: Dream, Mirage or Nemesis?* (Oxford: Blackwell, 1979).

11. D. St. George, "Life Expectancy, Truth, and the ABPI," *Lancet*, August 9, 1986, p. 346.
12. J. B. McKinlay, S. M. McKinlay, and R. Beaglehole, "Trends in Death and Disease and the Contribution of Medical Measures" in H. E. Freeman and S. Levine, eds., *Handbook of Medical Sociology* (Englewood Cliffs, N.J.: Prentice Hall, 1988), p. 16.
13. See William Paton, *Man and Mouse* (Oxford: Oxford University Press, 1984); Andrew Rowan, *Of Mice, Models and Men: A Critical Evaluation of Animal Research* (Albany: State University of New York Press, 1984), chap. 12; Michael DeBakey, "Medical Advances Resulting from Animal Research," in J. Archibald, J. Ditchfield, and H. Rowsell, eds., *The Contribution of Laboratory Animal Science to the Welfare of Man and Animals: Past, Present and Future* (New York: Gustav Fischer Verlag, 1985); OTA, *Alternatives to Animal Use in Research, Testing and Education*, chap. 5; and National Research Council, *Use of Animals in Biomedical and Behavioral Research* (National Academy Press, Washington, D.C., 1988), chap. 3.
14. Probably the best of those works arguing against the claims made for animal experimentation is Robert Sharpe, *The Cruel Deception* (Wellingborough, England: Thorsons, 1988).
15. "The Costs of AIDS," *New Scientist*, March 17, 1988, p. 22.

DOWN ON THE FACTORY FARM . . .

1. *Washington Post*, October 3, 1971; see also the testimony during September and October 1971, before the Subcommittee on Monopoly of the Select Committee on Small Business of the U.S. Senate, in the Hearings on the Role of Giant Corporations, especially the testimony of Jim Hightower of the Agribusiness Accountability Project. For the size of egg producers, see *Poultry Tribune*, June 1987, p. 27.
2. *Stall Street Journal*, July 1972.
3. J. Webster, C. Saville, and D. Welchman, "Improved Husbandry Systems for Veal Calves," Animal Health Trust and Farm Animal Care Trust, n.d., p. 5; see also Webster et al., "The Effect of Different Rearing Systems on the Development of Calf Behavior," and "Some Effects of Different Rearing Systems on Health, Cleanliness and Injury in Calves," *British Veterinary Journal* 1141: 249 and 472 (1985).
4. J. Webster, C. Saville, and D. Welchman, "Improved Husbandry Systems for Veal Calves," p. 6.
5. Ibid., p. 2.
6. *Stall Street Journal*, November 1973.
7. *Stall Street Journal*, April 1973.
8. *Stall Street Journal*, November 1973.
9. *Farmer and Stockbreeder*, September 13, 1960, quoted by Ruth Harrison, *Animal Machines* (London: Vincent Stuart, 1964) p. 70.
10. *Stall Street Journal*, April 1973.

11. G. van Putten, "Some General Remarks Concerning Farm Animal Welfare in Intensive Farming Systems," unpublished paper from the Research Institute for Animal Husbandry, "Schoonoord," Driebergseweg, Zeist, Netherlands, p. 2.
12. Ibid., p. 3.
13. *Vealer*, March/April 1982.
14. U.K. Ministry of Agriculture, Fisheries and Food, Welfare of Calves Regulations, 1987 (London: Her Majesty's Stationery Office, 1987).

BRIDGING THE GAP

1. John Locke's definition of a person is to be found in his *Essay on Human Understanding*, bk. II. chap. 27, paragraph 9.
2. The description comes from Frans de Waal's fascinating book, *Chimpanzee Politics* (London: Cape, 1982). I made only two changes: I gave human names to the chimpanzees involved in the fight, and I added from another study of chimpanzees the experiment involving the two series of boxes.
3. Charles Darwin, *Notebooks, 1836–1844*, Paul H. Barrett et al., eds. (Ithaca, N.Y.: Cornell University Press, 1987), p. 300.
4. Peter Singer, *Animal Liberation* (New York: New York Review of Books, 1975); 2nd ed. (1990). (See "All Animals Are Equal . . ." in this volume.) For an account of the history of the movement, see Richard Ryder, *Animal Revolution: Changing Attitudes Towards Speciesism* (Oxford: Blackwell, 1989).
5. See the essays on this topic in Paola Cavalieri and Peter Singer, eds., *The Great Ape Project* (London: Fourth Estate, 1993); and also E. Sue Savage-Rumbaugh, et al., *Apes, Language, and the Human Mind* (New York: Oxford University Press, 1998).
6. Jared Diamond, *The Rise and Fall of the Third Chimpanzee*, (New York: HarperCollins, 1991). The relevant passage is reprinted in *The Great Ape Project*, pp. 88–101. The preceding paragraph draws on Diamond's account.
7. Toshisada Nishida, "Chimpanzees Are Always New to Me" in *The Great Ape Project* (London: Fourth Estate, 1993), 24.
8. The Great Ape Project can be contacted at P.O. Box 19492, Portland, OR 97280-0492, USA, or through its web site, www.greatapeproject.org.

ENVIRONMENTAL VALUES

1. For more on the Franklin Dam, see the chapter "Living Ethically."
2. I have defended this assumption elsewhere: see the chapter "About Ethics."
3. Genesis 1:26–28.
4. See Robin Attfield, *The Ethics of Environmental Concern* (Oxford: Blackwell, 1983).
5. Genesis 9:2.

6. Corinthians 9:9–10.
7. Augustine, *The Catholic and Manichean Ways of Life*, D. A. Gallagher and I. J. Gallagher, trans. (Boston: Catholic University Press, 1966), p. 102. For cursing of the fig tree, see Mark 11:12–22, and for that of the pigs, Mark 5:1–13.
8. Aristotle, *Politics* (London: Dent, 1916), p. 16.
9. Aquinas, *Summa Theologica*, II, ii, question 64, article 1; I, ii, question 72, article 4.
10. For details on the alternative Christian thinkers, see Keith Thomas, *Man and the Natural World* (London: Allen Lane, 1983), pp. 152–153; and Attfield, op. cit (note 4).
11. Aristotle, op. cit. (note 8).
12. Albert Schweitzer, *Civilization and Ethics* (part II of *The Philosophy of Civilization*, C. T. Campion, trans., 2nd ed. (London: A. & C. Black, 1929), pp. 246–7.
13. Paul Taylor, *Respect for Nature* (Princeton, N.J.: Princeton University Press, 1986), pp. 45, 128. My discussion draws on a fine critique of Taylor by Gerald Paske, "The Life Principle: A (metaethical) rejection," *Journal of Applied Philosophy* 6: 219–225 (1989).
14. A. Leopold, *A Sand Country Almanac* (New York: Oxford University Press, 1966; first published 1949), pp. 219, 238.
15. A. Naess, "The Shallow and the Deep, Long-Range Ecology Movement," *Inquiry* 16: 95–100 (1973).
16. See, for example, the following works: W. Devall and G. Sessions, *Deep Ecology: Living As If Nature Mattered*, (Salt Lake City: Gibbs Smith, 1985); L. Johnson, *A Morally Deep World* (Cambridge: Cambridge University Press, 1990); V. Plumwood, "Ecofeminism: an Overview and Discussion of Positions and Arguments: Critical Review," *Australasian Journal of Philosophy* 64 (supplement), 120–138 (1986); R. Sylvan, "Three Essays Upon Deeper Environmental Ethics," *Discussion Papers in Environmental Philosophy* 13 (1986), published by the Australian National University, Canberra; and P. Taylor, op. cit.
17. Leopold, op. cit., p. 262.
18. A. Naess and G. Sessions, "Basic Principles of Deep Ecology," *Ecophilosophy* 6: 3–7 (1984), quoted from D. Bennet and R. Sylvan, "Australian Perspectives on Environmental Ethics: A UNESCO Project" (unpublished, 1989).
19. R. Routley [now R. Sylvan] and V. Routley [now V. Plumwood], "Human Chauvinism and Environmental Ethics" in D. Mannison, M. McRobbie, and R. Routley, eds., *Environmental Philosophy* (Canberra: Australian National University Research School of Social Sciences, 1980).
20. For a useful survey of the value positions of deep ecologists, see R. Sylvan, *A Critique of Deep Ecology* (Canberra: Australian National University, Department of Philosophy, Discussion Papers in Environmental Philosophy, 12, 1985), p. 53.

FAMINE, AFFLUENCE, AND MORALITY

1. There was also a third possibility: that India would go to war to enable the refugees to return to their lands. Since I wrote this paper, India has taken this way out. The situation is no longer that described above, but this does not affect my argument, as the next paragraph indicates.
2. In view of the special sense philosophers often give to the term, I should say that I use "obligation" simply as the abstract noun derived from "ought," so that "I have an obligation to" means no more, and no less, than "I ought to." This usage is in accordance with the definition of "ought" given by the *Shorter Oxford English Dictionary*: "the general verb to express duty or obligation." I do not think any issue of substance hangs on the way the term is used; sentences in which I use "obligation" could all be rewritten, although somewhat clumsily, as sentences in which a clause containing "ought" replaces the term "obligation."
3. J. O. Urmson, "Saints and Heroes," in *Essays in Moral Philosophy*, Abraham I. Melden, ed. (Seattle and London: University of Washington Press, 1958), p. 214. For a related but significantly different view see also Henry Sidgwick, *The Methods of Ethics*, 7th ed. (London: Macmillan, 1907), pp. 220–221; 492–493.
4. *Summa Theologica*, II-II, question 66, article 7, in *Aquinas, Selected Political Writings*, A. P. d'Entreves, ed., J. G. Dawson, trans. (Oxford: Clarendon Press, 1948), p. 171.
5. See, for instance, John Kenneth Galbraith, *The New Industrial State* (Boston: Houghton Mifflin, 1967); and E. J. Mishan, *The Costs of Economic Growth* (New York: Prager, 1967).

WHAT'S WRONG WITH KILLING?

1. Andrew Stinson's treatment is described by Robert and Peggy Stinson in *The Long Dying of Baby Andrew* (Boston: Little, Brown, 1983).
2. For more on this subject, see "All Animals Are Equal," pp. 28–46 in this volume.
3. Joseph Fletcher's article "Indicators of Humanhood: A Tentative Profile of Man" appeared in *Hastings Center Report*, 2, 5 (1972).
4. John Locke's definition of "person" is taken from his *Essay Concerning Human Understanding*, bk. II, chap. 27, paragraph 9.
5. Aristotle's views on infanticide are in his *Politics*, bk. 7, p. 1335b; Plato's are in the *Republic*, bk. 5, p. 460.
6. Support for the claim that our present attitudes toward infanticide are largely the effect of the influence of Christianity on our thought can be found in the historical material on infanticide cited in note 9 in "Taking Life: The Embryo and the Fetus." (See especially the article by W. L. Langer, pp. 353–355.)

7. For Aquinas's statement that killing a human being offends against God as killing a slave offends against the master of the slave, see *Summa Theologica*, 2, ii, question 64, article 5.

8. Hare propounds and defends his two-level view of moral reasoning in *Moral Thinking* (Oxford: Clarendon Press, 1981).

9. Michael Tooley's "Abortion and Infanticide" was first published in *Philosophy and Public Affairs* 2 (1972). The passage quoted here is from a revised version in J. Feinberg, ed., *The Problem of Abortion* (Belmont, California: Wadsworth, 1973), p. 60. His book *Abortion and Infanticide* was published by Clarendon Press in Oxford in 1983.

10. For further discussion of respect for autonomy as an objection to killing, see Jonathan Glover, *Causing Death and Saving Lives* (Harmondsworth, Middlesex, England: Penguin, 1977), chap. 5; and H. J. McCloskey, "The Right to Life," *Mind* 84 (1975).

11. My discussion of the "total" and "prior existence" versions of utilitarianism owes much to Derek Parfit. I originally tried to defend the prior existence view in "A Utilitarian Population Principle," in M. Bayles, ed., *Ethics and Population* (Cambridge, Mass.: Schenkman, 1976), but Parfit's reply, "On Doing the Best for Our Children," in the same volume, persuaded me to change my mind. Parfit's *Reasons and Persons* (Oxford: Clarendon Press, 1984) is required reading for anyone wishing to pursue this topic in depth. See also his short account of some of the issues in "Overpopulation and the Quality of Life," in P. Singer, ed., *Applied Ethics* (Oxford: Oxford University Press, 1986). Parfit uses the term "person-affecting" where I use "prior existence." The reason for the change is that the view has no special reference to persons, as distinct from other sentient creatures. The distinction between the two versions of utilitarianism appears to have been first noticed by Henry Sidgwick, *The Methods of Ethics* (London: Macmillan, 1907), pp. 414–416. Later discussions include, in addition to those cited above, J. Narveson, "Moral Problems of Population," *Monist* 57 (1973); T. G. Roupas, "The Value of Life," *Philosophy and Public Affairs* 7 (1978); and R. I. Sikora, "Is It Wrong to Prevent the Existence of Future Generations?" in B. Barry and R. Sikora, eds., *Obligations to Future Generations* (Philadelphia: Temple University Press, 1978).

12. Mill's famous passage comparing Socrates and the fool appeared in his *Utilitarianism* (London: J. M. Dent, 1960; first published 1863), pp. 8–9.

TAKING LIFE: THE EMBRYO AND THE FETUS

1. The most important sections of the decision of the U.S. Supreme Court in *Roe v. Wade* are reprinted in J. Feinberg, ed., *The Problem of Abortion* (Belmont, California: Wadsworth, 1973).

2. The government committee referred to in the subsection "Not the Law's Business?"—the Wolfenden Committee—issued the *Report of the Committee on Homosexual Offences and Prostitution*, Command Paper 247 (London: Her Majesty's Stationery Office, 1957), p. 24.

3. J. S. Mill's "very simple principle" is stated in the introductory chapter of *On Liberty* (London: J. M. Dent, 1960), p. 73.

4. Edwin Schur's *Crimes Without Victims* was published by Prentice-Hall in Englewood Cliffs, N.J., in 1965.
5. Judith Jarvis Thomson's "A Defense of Abortion" appeared in *Philosophy and Public Affair*, 1 (1971) and has been reprinted in P. Singer, ed., *Applied Ethics* (Oxford: Oxford University Press, 1986).
6. Paul Ramsey uses the genetic uniqueness of the fetus as an argument against abortion in "The Morality of Abortion," in D. H. Labby, ed., *Life or Death: Ethics and Options* (London: University of Washington Press, 1968); reprinted in J. Rachels, ed., *Moral Problems*, 2nd ed. (New York: Harper & Row, 1975), p. 40.
7. Bentham's reassuring comment on infanticide, quoted in the section "Abortion and Infanticide" is from his *Theory of Legislation* (London: Paul, Trench, Trubner, 1931), and is quoted by E. Westermarck, *The Origin and Development of Moral Ideas* (London: Macmillan, 1924), vol. 1, p. 413n.
8. In the final part of *Abortion and Infanticide* (Oxford: Clarendon Press), Michael Tooley discusses the available evidence on the development in the infant of the sense of being a continuing self.
9. For historical material on the prevalence of infanticide, see Maria Piers, *Infanticide* (New York, 1978); and W. L. Langer, "Infanticide: A Historical Survey," *History of Childhood Quarterly* 1 (1974). An older but still valuable survey is in Edward Westermarck, *The Origin and Development of Moral Ideas*, vol. 1, (London: Macmillan, 1924), pp. 394–413. An interesting study of the use of infanticide as a form of family planning is *Nakahara: Family Farming and Population in a Japanese Village, 1717–1830*, by Thomas C. Smith (Palo Alto, Calif.: Stanford University Press, 1977).
10. References for Plato and Aristotle were given in the notes to "What's Wrong With Killing?"
11. For Seneca, see *De Ira* 1, 15, cited by Westermarck, *The Origin and Development of Moral Ideas*, vol. 1, p. 419.

PROLOGUE

1. Don Colburn, "AMA Ethics Panel Revises Rules on Withholding Food; In Irreversible Comas, Water and Nutrition May Be Stopped," *Washington Post*, April 2, 1986, p. 9.
2. Stuart Youngner et al., " 'Brain Death' and Organ Retrieval: A Cross-Sectional Survey of Knowledge and Concepts Among Health Professionals," *Journal of the American Medical Association*, 261: 2209 (1989).
3. Carole Outterson, "Newborn Infants with Severe Defects: A Survey of Paediatric Attitudes and Practices in the UK," *Bioethics* 7: 420–435 (1993); see questions 7, 16, and 17, but the percentage of respondents who accepted all of these propositions is not apparent from the published article. I am grateful to Ms. Outterson for making it available to me.
4. Eric Harrison and Tracy Shryer, "Weeping Father Pulls Gun, Stops Infant's Life Support," *Los Angeles Times*, April 27, 1989, p. 1.

IS THE SANCTITY OF LIFE ETHIC TERMINALLY ILL?

1. Henry Beecher to Robert Ebert, 30 October 1967. The letter is in the Henry Beecher Manuscripts at the Francis A. Countway Library of Medicine, Harvard University, and is quoted by David Rothman, *Strangers at the Bedside* (New York: Basic Books, 1991), pp. 160–161.
2. The first draft and Ebert's comment on it are both quoted by Rothman, *Strangers at the Bedside*, pp. 162–164. The documents are in the Beecher Manuscript collection.
3. Henry Beecher, "The New Definition of Death, Some Opposing Viewpoints," *International Journal of Clinical Pharmacology* 5: 120–121 (1971) (italics in original).
4. President's Commission for the Study of Ethical Problems in Medicine, *Defining Death: A Report on the Medical, Legal and Ethical Issues in the Determination of Death* (Washington, D.C.: U.S. Government Printing Office, 1981), pp. 24, 25.
5. Ibid., pp. 67, 72.
6. In 1998 Japan adopted a law allowing organs to be taken from brain-dead patients.
7. Stuart Youngner et al., " 'Brain Death' and Organ Retrieval: A Cross-Sectional Survey of Knowledge and Concepts Among Health Professionals," *Journal of the American Medical Association*, 261 (1090): 2209 (1989).
8. See, for example, the United States Uniform Determination of Death Act. Note that the Harvard committee had referred to the absence of central nervous system "activity" rather than function. The use of the term "function" rather than "activity" makes the definition of brain death more permissive, because, as the United States President's Commission recognized (*Defining Death*, p. 74), electrical and metabolic activity may continue in cells or groups of cells after the organ has ceased to function. The commission did not think that the continuation of this activity should prevent a declaration of death.
9. Robert Truog, "Rethinking Brain Death," in K. Sanders and B. Moore, eds., *Anencephalics, Infants and Brain Death Treatment Options and the Issue of Organ Donation* (Melbourne: Law Reform Commission of Victoria, 1991), pp. 62–74; Amir Halevy and Baruch Brody, "Brain Death: Reconciling Definitions, Criteria and Tests," *Annals of Internal Medicine* 119 (6): 519–525 (1993); Robert Veatch, "The Impending Collapse of the Whole-Brain Definition of Death," *Hastings Center Report* 23 (4): 18–24 (1993).
10. Henry Beecher, "The New Definition of Death, Some Opposing Views," unpublished paper presented at the meeting of the American Association for the Advancement of Science, December 1970, p. 4, quoted from Robert Veatch, *Death, Dying and the Biological Revolution* (New Haven, Conn.: Yale University Press, 1976), p. 39.

11. *Cruzan* v. *Director, Missouri Department of Health* (1990) 110 S. Ct. pp. 2886–7.
12. *Airedale N.H.S. Trust* v. *Bland (C.A)* (February 19, 1993) 2 *Weekly Law Reports*, p. 350. Page numbers given without further identifying details in subsequent footnotes are to this report of the case.
13. P. 333; the passage was quoted again by Lord Goff of Chieveley in his judgment in the House of Lords, p. 364.
14. Pp. 374, 386.
15. P. 331.
16. John Keown, "Courting Euthanasia? Tony Bland and the Law Lords," *Ethics and Medicine* 9 (3): 36 (1993).
17. P. 339.
18. P. 361.
19. P. 400.
20. P. 383.
21. P. 388.
22. *R.* v. *Adams* (1959), quoted by Derek Morgan, "Letting Babies Die Legally," *Institute of Medical Ethics Bulletin*, May 1989, p. 13. See also Patrick Devlin, *Easing the Passing: The Trial of Dr. John Bodkin Adams* (London: Faber and Faber, 1986), pp. 171, 209.
23. Pp. 388–89.
24. For a statement of the traditional definition, see, for example, *Black's Law Dictionary* (New York: West, 1968).

JUSTIFYING VOLUNTARY EUTHANASIA

1. On euthanasia in the Netherlands, see J. K. Gevers, "Legal Developments Concerning Active Euthanasia on Request in the Netherlands," *Bioethics* 1 (1987). The annual number of cases is given in "Dutch Doctors Call for Legal Euthanasia," *New Scientist*, October 12, 1991, p. 17. Paul J. van der Maas et al., "Euthanasia and Other Medical Decisions Concerning the End of Life," *Lancet*, 338: 669–674 (September 14, 1991), at 673, gives a figure of 1900 deaths due to euthanasia each year, but this is limited to reports from doctors in general practice. The quotation about patients' desire for reassurance comes from this article, p. 673.
2. The case of Diane is cited from Timothy E. Quill, "Death and Dignity: A Case of Individualized Decision Making," *New England Journal of Medicine* 324 (10): 691–694 (March 7, 1991). Betty Rollin describes the death of her mother in Betty Rollin, *Last Wish* (New York: Penguin, 1987). The passage quoted is from pp. 149–150. See also Betty Rollin's foreword to Derek Humphry, *Final Exit: The Practicalities of Self-Deliverance and Assisted Suicide* (Eugene, Ore., 1991), pp. 12–13.

EUTHANASIA: EMERGING FROM HITLER'S SHADOW

1. Leo Alexander, "Medical Science under Dictatorship," *New England Journal of Medicine* 241: 39–47 (July 14, 1949).
2. Michael Burleigh, *Death and Deliverance: "Euthanasia" in Germany, c. 1900–1945* (Cambridge: Cambridge University Press, 1994), pp. 3–4.
3. Arlene Judith Klotzko, "What Kind of Life? What Kind of Death? An Interview with Dr. Henk Prins," *Bioethics* 11 (1): 24–42 (January 1997).
4. I am grateful to Dr Dörner for sending me a copy of his unpublished lecture. The passage I have quoted is on p. 7. For a fuller presentation of his views on the origins of the Nazi crimes, see his *Tödliches Mitleid*, 3rd ed. (Gütersloh: Verlag Jakob von Hoddis, 1993).
5. For further details see my *Rethinking Life and Death* (Melbourne: Text, 1994; Oxford: Oxford University Press and New York: St. Martin's Press, 1995).
6. Christian Munthe, *The Moral Roots of Prenatal Diagnosis*, (Göteborg: Studies in Research Ethics No. 7, Centre for Research Ethics, 1996), p. 46n.
7. See Helga Kuhse and Peter Singer, *Should the Baby Live?* (Oxford: Oxford University Press, 1985), pp. 11–17.
8. Klotzko, op. cit., p. 39.
9. Ibid., p. 31.

IN PLACE OF THE OLD ETHIC

1. The classic account of the shift from the Ptolemaic to the Copernican model is Thomas Kuhn, *The Structure of Scientific Revolutions* (Chicago, Ill.: University of Chicago Press, 1972).
2. Dr. L. Haas, from a letter in *Lancet*, November 2, 1968; quoted from S. Gorovitz, ed., *Moral Problems in Medicine* (Englewood Cliffs, N.J.: Prentice-Hall, 1976), p. 351.
3. Clough's "The Latest Decalogue" can be found in Helen Gardner, ed., *The New Oxford Book of English Verse* (Oxford: Oxford University Press, 1978).
4. Thomas Aquinas, *Summa Theologica*, II, ii, question 64, article 5.
5. John Stuart Mill, *On Liberty* (London: Dent, 1960), pp. 72–73.
6. This position is associated with Michael Tooley's influential article, "Abortion and Infanticide," *Philosophy and Public Affairs* 2: 37–65 (1972); for a slightly different argument to the same conclusion, see also Michael Tooley, *Abortion and Infanticide* (Oxford: Oxford University Press, 1983). Similar views have been defended by several philosophers and bioethicists, among them H. Tristram Engelhardt, Jr., *The Foundations of Bioethics* (New York: Oxford University Press, 1986); R. G. Frey, *Rights, Killing and Suffering* (Oxford: Blackwell, 1983); Jonathan Glover, *Causing Death and Saving Lives* (Harmondsworth: Penguin, 1977); John Harris, *The Value of Life* (London: Routledge and Kegan Paul, 1985); Helga Kuhse, *The Sanctity of Life Doctrine in Medicine: A Critique* (Ox-

ford: Oxford University Press, 1987); James Rachels, *The End of Life* (Oxford: Oxford University Press, 1986); and *Created from Animals* (Oxford: Oxford University Press, 1991). See also my own *Practical Ethics*, 2nd ed. (Cambridge: Cambridge University Press, 1993; originally published 1979).

7. For Augustine, see *Against Faustus*, bk. 15. chap. 7; for Luther, "Der Grosse Catechismus," 1529, "On the Sixth Commandment"; and for Calvin, *Commentaries on the First Book of Moses Called Genesis*, vol. 2, chap. 38, v. 8. I owe these references and other information in this paragraph to John T. Noonan, "Contraception," in Warren T. Reich, ed., *Encyclopedia of Bioethics* (New York: Free Press, 1978), vol. I, pp. 204–216. The Supreme Court case referred to is *Griswold* v. *Connecticut*, 1965.
8. Jodi L. Jacobson, "Holding Back the Sea," in Lester Brown et al., *State of the World, 1990: The Worldwatch Institute Report on Progress Towards a Sustainable Economy* (Washington, D.C.: Worldwatch Institute, 1990).
9. Peter Singer, "Sanctity of Life or Quality of Life," *Pediatrics* 72: 128–129 (July 1983); three protest letters were published with my reply in 73: 259–263 (February 1984), but the remainder of the letters are unpublished.
10. Albert Schweitzer's ethic of reverence for life *may* be making this wider claim; the contemporary American philosopher Paul Taylor certainly does make it. See "Environmental Values," pp. 86–102.
11. For an argument that it could be as late as 32 weeks, see Susan Taiwa, "When Is the Capacity for Sentience Acquired During Human Fetal Development?" *Journal of Maternal-Fetal Medicine* 1: 153–165 (1992).
12. For a full defense of this position, see Helga Kuhse and Peter Singer, *Should the Baby Live?* (Oxford: Oxford University Press, 1985).
13. Shakespeare, *Timon of Athens*, Act 5, Scene 1.
14. For sources and further details, see the notes to "Taking Life: The Embryo and the Fetus."
15. Robert and Peggy Stinson, *The Long Dying of Baby Andrew* (Boston: Little, Brown, 1983), pp. 153, 266–267.
16. Here I have been influenced by Norbert Hoerster, "Kindestötung und das Lebensrecht von Personen," *Analyse & Kritik* 12: 226–244 (1990).
17. Thomas Hobbes, *Leviathan*, chap. 13.

THE ULTIMATE CHOICE

1. Information in this and the following paragraphs on Ivan F. Boesky is taken in part from Robert Slater, *The Titans of Takeover* (Englewood Cliffs, N.J.: Prentice-Hall, 1987), chap. 7.
2. *Wall Street Journal*, June 20, 1985; quoted in Slater, p. 134.
3. Ivan F. Boesky, *Merger Mania* (New York: Holt, Rinehart and Winston, 1985), p. v. The earlier quotations are from pp. xiii–xiv.

4. Mark Brandon Read, *Chopper from the Inside* (Kilmore, Vic: Floradale Productions, 1991), pp. 6–7.
5. Plato, *Republic*, bk. II, 360, 2nd ed., Desmond Lee, trans. (Harmondsworth, Middlesex, England: Penguin, 1984).
6. Robert J. Ringer, *Looking Out for # 1* (New York: Fawcett Crest, 1978), p. 22.
7. Todd Gitlin, *Inside Prime Time* (New York: Pantheon, 1983), pp. 268–269.

LIVING ETHICALLY

1. For details on Wallenberg's life, see John Bierman, *The Righteous Gentile* (New York: Viking, 1981).
2. See Thomas Kenneally, *Schindler's Ark* (London: Hodder and Stoughton, 1982).
3. Samuel and Pearl Oliner, *The Altruistic Personality: Rescuers of Jews in Nazi Europe* (New York: Free Press, 1988). The cases mentioned earlier in the paragraph are taken from Kristen R. Monroe, Michael C. Barton, and Ute Klingemann, "Altruism and the Theory of Rational Action: Rescuers of Jews in Nazi Europe," *Ethics* 101 (1): 103–123 (October 1990). See also Perry London, "The Rescuers: Motivational Hypotheses About Christians Who Saved Jews from the Nazis," in J. Macaulay and L. Berkowitz, eds., *Altruism and Helping Behavior* (New York: Academic, 1970); Carol Rittner and Gordon Myers, eds., *The Courage to Care—Rescuers of Jews During the Holocaust* (New York: New York University Press, 1986); Nehama Tec, *When Light Pierced the Darkness—Christian Rescuers of Jews in Nazi-Occupied Poland* (New York: Oxford University Press, 1986); and Gay Block and Malka Drucker, *Rescuers—Portraits of Moral Courage in the Holocaust* (New York: Holmes and Meier, 1992).
4. Primo Levi, *If This Is a Man*, Stuart Woolf, trans. (London: Abacus, 1987), pp. 125, 127–128.
5. The story of Corti and Delaney is the subject of Jonathan Kwitny's *Acceptable Risks* (New York: Poseidon, 1992).
6. The Blockaders, *The Franklin Blockade* (Hobart: Wilderness Society, 1983), p. 72.
7. *Conservation News* 24, 2 (April/May 1992).
8. Maimonides, *Mishneh Torah*, bk. 7, chap. 10, reprinted in Isadore Twersky, *A Maimonides Reader* (New York: Behrman House, 1972), pp. 136–137.
9. R. M. Titmuss, *The Gift Relationship* (London: Allen and Unwin, 1971), p. 44.
10. These figures were obtained from correspondence received from the relevant bone marrow registries during June–July 1992.
11. Alfie Kohn, *The Brighter Side of Human Nature* (New York: Basic Books, 1990), p. 64.
12. B. O'Connell, "Already 1,000 Points of Light," *New York Times*, Jan. 25, 1989, A23. (I owe this reference to Alfie Kohn, *The Brighter Side of Human Nature*, p. 290.) See also *Time*, April 8, 1991.

13. Production of aerosol-powered personal care products in 1989 declined 11 percent from 1988 levels, according to the Chemical Specialty Manufacturers Association, *The Rose Sheet* (Chevy Chase, Md.), Dec. 10, 1990. Vol. 11, No. 50.
14. "Doing the Right Thing," *Newsweek*, Jan. 7, 1991, pp. 42–43.
15. The quotations are taken from Titmuss, *The Gift Relationship*, pp. 227–228.
16. E. Lightman, "Continuity in Social Policy Behaviors: The Case of Voluntary Blood Donorship," *Journal of Social Policy* 10 (1): 53–79 (1981); J. A. Piliavin, D. E. Evans, and P. Callero, "Learning to 'Give to unnamed strangers': The Process of Commitment to Regular Blood Donation," in E. Staub et al., eds., *Development and Maintenance of Prosocial Behavior: International Perspectives on Positive Morality* (New York: Plenum, 1984), pp. 471–491; J. Piliavin, "Why Do They Give the Gift of Life? A Review of Research on Blood Donors Since 1977," *Transfusion* 30 (5): 444–459 (1990). For Aristotle's views on virtue, see his *Nicomachean Ethics*, W. D. Ross, trans. (London: World Classics, Oxford University Press, 1959). I take the point made in this paragraph from "Giving Blood: The Development of Generosity," unsigned article in *Issues in Ethics* 5, 1 (1992), published by the Santa Clara University Center for Applied Ethics, Calif.

THE GOOD LIFE

1. I used the metaphor of "the escalator of reason" in my book *The Expanding Circle*, p. 88; some parts of this section draw on that work. Colin McGinn made essentially the same argument in "Evolution, Animals and the Basis of Morality," *Inquiry* 22: 91 (1979).
2. John Aubrey, *Brief Lives*, A. Clark, ed. (Oxford: Oxford University Press, 1898), vol. 1, p. 332.
3. Gunnar Myrdal, *An American Dilemma* (New York: Harper & Row, 1944), app. 1.
4. See the writings included in "Across the Species Barrier" in this volume.
5. Karl Marx, *The German Ideology* (New York: International, 1966), pp. 40–41.
6. For more detailed discussion of these points, see my book *Practical Ethics*, 2nd ed., pp. 232–234.
7. The sources are, respectively: *Gospel According to St. Matthew*, 22:39; *Babylonian Talmud*, Order Mo'ed, Tractate Sabbath, sec. 31a; Lun Yu XV:23 and XII:2, quoted by E. Westermarck, *The Origin and Development of the Moral Ideas* (London: Macmillan, 1924), vol. I, p. 102; and *Mahabharata*, XXIII:5571.
8. Luke, 10:29–37.

A MEANINGFUL LIFE

1. With a lot of help from others, especially John Swindells, who became coproducer and director, the video was eventually transformed into *Henry: One*

Man's Way, a documentary shown on SBS-TV, Australia, on August 22, 1997. The video is available in the United States through Bullfrog Films, Oley, Pa. (1-800-543-3674).

2. Joan Zacharias, "The Satya Interview: Making a Difference: An Interview with Henry Spira," *Satya*, July 1995, p. 9.
3. Henry is referring to the following passage from Barnaby Feder, "Pressuring Perdue," *New York Times Magazine*, November 26, 1989, p. 72: "When asked what his epitaph should be, he ponders and suggests, 'He pushed the peanut forward.' "

ON BEING SILENCED IN GERMANY

1. Cambridge: Cambridge University Press, 1979; German translation, *Praktische Ethik* (Stuttgart: Reclam, 1984); Spanish translation, *Etica Practica* (Barcelona: Ariel, 1984); Italian translation, *Etica Praticà* (Naples: Liguori, 1989); Swedish translation, *Praktisk Ethik* (Stockholm: Thales, 1990).
2. Oxford: Oxford University Press/Clarendon Press, 1987.
3. *Der Standard* (Vienna), October 10, 1990.
4. *Eine Frage des Lebens: Ethik der Abtreibung and Künstlichen Befruchtung* (Frankfurt: Campus, 1990).
5. *Analyse & Kritik*, December 12, 1990.
6. During the period when opposition to the Wittgenstein Symposium was being stirred up, these philosophers were all described, in terms calculated to arouse a hostile response, in a special "euthanasia issue" of the Austrian journal *erziehung heute* (*education today*) (Innsbruck, 1991), p. 37.
7. Adolf Hübner, "Euthanasie diskussion im Geiste Ludwig Wittgenstein?" *Der Standard* (Vienna), May 21, 1991.
8. "Die krisenhafte Situation der Österreichischen Ludwig Wittgenstein Gesellschaft, ausgelöst durch die Einladungspraxis zum Thema 'Angewandte Ethik' " (unpublished typescript).
9. Martin Stürzinger, "Ein Tötungshelfer mit faschistischem Gedankengut?" *Die Weltwoche* (Zurich), May 23, 1991, p. 83.
10. There is a brief account of my reasons for holding this position in *Practical Ethics*, chap. 7; and a much more detailed one in Helga Kuhse and Peter Singer, *Should the Baby Live?* (Oxford: Oxford University Press, 1985). See also Peter Singer and Helga Kuhse, "The Future of Baby Doe," *New York Review* (March 1, 1984), pp. 17–22.
11. Here is a selection; many more could be added: H. Tristram Engelhardt, Jr., *The Foundations of Bioethics* (Oxford: Oxford University Press, 1986); R. G. Frey, *Rights, Killing and Suffering* (London: Blackwell, 1983); Jonathan Glover, *Causing Deaths and Saving Lives* (New York: Penguin, 1977); John Harris, *The Value of Life* (London: Routledge and Kegan Paul, 1985); James Rachels, *The End of Life* (Oxford: Oxford University Press, 1986); and *Created from Animals* (Oxford:

Oxford University Press, 1991); Michael Tooley, *Abortion and Infanticide* (Oxford: Oxford University Press, 1983); and the book by Helga Kuhse to which I have already referred, *The Sanctity-of-Life Doctrine in Medicine: A Critique.*

12. Franz Christoph, "(K)ein Diskurs über 'lebensunwertes Leben,'" *Der Spiegel* (23) (June 5, 1989).

13. "Bizarre Verquickung" and "Wenn Mitleid tödlich wird," *Der Spiegel* (34): 171–176 (August 21, 1989).

14. My *Animal Liberation* (New York: Random House, 1975; 2nd rev. ed., New York: New York Review/Random House, 1990) had been published in Germany under the title *Befreiung der Tiere* (Munich: F. Hirthammer, 1982) but it was not widely known at the time. Nevertheless, *Practical Ethics* contains two chapters summarizing my views on animals, so the response did indicate that most of the protesters had not read the book on which they based their opposition to my invitation to speak.

15. For this reason one of the protesters, reporting on the events in a student publication, made it clear that to enter into the discussion with me was a tactical error. See Holger Dorff, "Singer in Saarbrücken," *Unirevue* (Winter semester, 1989/1990), p. 47.

16. Helga Kuhse, "Warum Fragen der Euthanasie auch in Deutschland unvermeidlich sind," *Deutsche Ärzteblatt* (16): 1243–9 (April 19, 1990); readers' letters, and a response by Kuhse, are to be found in (37): 2696–704 (September 13, 1990), and (38): 2792–6 (September 20, 1990).

17. The list of books published between January 1990 and September 1991 devoted to this theme includes: C. Anstötz, *Ethik und Behinderung* (Berlin: Edition Marhold, 1990); T. Bastian, ed., *Denken, Schreiben, Töten* (Stuttgart: Hirzel, 1990); T. Bruns, U. Panselin, and U. Sierck, *Tödliche Ethik* (Hamburg: Verlag Libertäre Assoziation, 1990); Franz Christoph, *Tödlicher Zeitgeist* (Cologne: Kiepenheuer und Witsch, 1990); R. Hegselmann and R. Merkel, eds., *Zur Debatte über Euthanasie* (Frankfurt: Suhrkamp, 1991); E. Klee, *Durch Zyankali Erlöst* (Frankfurt: Fischer, 1990); A. Leist, ed., *Um Leben und Tod* (Frankfurt: Suhrkamp, 1990), and O. Tolmein, *Geschätzles Leben* (Hamburg: Konkret Literatur Verlag, 1990).

18. See, for instance, the way in which Rudi Tarneden, a reviewer from an association for the disabled, and very sympathetic to Christoph's position, is drawn in the course of his review to raise such questions as: "Aren't there in fact extreme situations of human suffering, limits to what is bearable? Am I really guilty of contempt for humanity [*Menschenverachtung,* a term often used in Germany to describe what I am supposed to be guilty of—P.S.] if I try to take this into account?" Rudi Tarneden, "Wo alles richtig ist, kann es auch keine Schuld mehr geben" (review of Franz Christoph, *Tödlicher Zeitgeist* and Christoph Anstötz, *Ethik und Behinderung*), *Zeitschrift für Heilpädagogik* 42 (4): 246 (1991).

19. In 1993 Rowahlt, a major German publisher, agreed to publish a German edition. When it was listed as "forthcoming" in the catalogue, the publisher received protests and threats, to which it eventually yielded, canceling the

planned publication. The book was subsequently published in Germany by Harold Fischer Verlag of Erlangan.

20. *Taz* (Berlin), January 10, 1990.

21. German feminists who read Franz Christoph's recent book (see note 17) may reconsider their support for his position; for he leaves no doubt that he is opposed to granting women a right to decide about abortion. For Christoph, "Abortion decisions are always decisions about whether a life is worthy of being lived; the child does not fit into the woman's present life-plans. Or: the social situation is unsatisfactory. Or: the woman holds that she is only able to bear a healthy child. Whether one likes it or not: with the last example, the woman who wants an abortion confirms an objectively negative social value judgment against the handicapped" (p. 13). There is more along these lines, all in a style well suited for quotation in the pamphlets of the antiabortion movement.

    This is, at least, more honest than the evasive maneuvering of Oliver Tolmein, who states in the foreword to his *Geschätztes Leben* that to discuss the significance of the feminist concept of self-determination in the context of prenatal diagnosis and abortion would take him "by far" beyond the bounds of his theme (p. 9). Odd, since the crux of his vitriolic attack on all who advocate euthanasia (an attack that includes, on the very first page of the book, a statement that it is necessary to disrupt seminars on the issue) is that those who advocate euthanasia are committed to valuing some human lives as not worth living.

22. R. M. Hare makes a similar point in a letter published in *Die Zeit*, August 11, 1989.

23. "Zur Sprengung einer Vortragsveranstaltung an der Universität," *Unipresse Dienst*, Universität Zurich, May 31, 1991.

24. See, for example, "Mit Trillerpfeifen gegen einen Philosophen," and "Diese Probleme kann and soll man besprechen," both in *Tages-Anzeiger*, May 29, 1991; "Niedergeschrien," *Neue Zürcher Zeitung*, May 27, 1991; and (despite the pejorative headline) "Ein Tötungshelfer mit faschistischem Gedankengut?" *Die Weltwoche*, May 23, 1991.

Permissions

"Moral Experts" by Peter Singer. Copyright © Peter Singer. From *Analysis*, vol. 32 (1972), pp. 115–117. Reprinted with permission.

"About Ethics," "What's Wrong with Killing?" "Taking Life: The Embryo and the Fetus," "Justifying Infanticide," and "Justifying Voluntary Euthanasia" by Peter Singer. Copyright © Peter Singer. From *Practical Ethics* (Cambridge: Cambridge University Press, 2nd ed., 1993). Reprinted with permission of Cambridge University Press.

"Preface to the 1975 Edition," "All Animals Are Equal," "Tools for Research," and "Down on the Factory Farm . . ." by Peter Singer. Copyright © Peter Singer. From *Animal Liberation* (New York: New York Review of Books/ Random House, 1975). Reprinted by permission of the author.

"A Vegetarian Philosophy" by Peter Singer. Copyright © Peter Singer. From *Consuming Passions: Food in the Age of Anxiety* (Manchester, England: Manchester University Press, 1998), pp. 72–75, 77–80. Reprinted with permission.

Passages from "Bridging the Gap" have appeared in *Rethinking Life and Death: The Collapse of Our Traditional Ethics* (Melbourne: Text Publishing, 1994; St. Martin's Press, New York, 1995) and in "The Rights of Ape," *BBC Wildlife*, June 1993, pp. 28–32.

"Environmental Values" by Peter Singer. Copyright © Peter Singer. From *The Environmental Challenge* (Melbourne: Lanman Cheshire, 1991), pp. 3–24. Reprinted with permission.

"Famine, Affluence, and Morality" by Peter Singer. From *Philosophy and Public Affairs* (Princeton, N.J.: Princeton University Press, Spring 1972, vol. 1, pp. 229–243). Copyright © 1972 by Princeton University Press. Reprinted by permission of Princeton University Press.

"The Singer Solution to World Poverty" by Peter Singer. Copyright © 1999 by Peter Singer. *New York Times Magazine,* September 5, 1999, pp. 60–63. Reprinted by permission of the author.

"Prologue" and "In Place of the Old Ethic" by Peter Singer. Copyright © 1995 by Peter Singer. From *Rethinking Life and Death: The Collapse of Our Traditional Values.* Reprinted by permission of St. Martin's Press, LLC.

"Is the Sanctity of Life Ethic Terminally Ill?" from *Bioethics,* vol. 9, no. 3, 4 (1995), pp. 327–343. (Boston: Blackwell Publishers, 1995). Reprinted with permission.

"The Ultimate Choice," "Living Ethically," and "The Good Life" by Peter Singer. From *How Are We to Live?* by Peter Singer (Amherst, N.Y.: Prometheus Books). Copyright © 1995. Reprinted by permission of the publisher.

"Darwin for the Left" by Peter Singer. First published in *Prospect* magazine, Britain's politics and current affairs monthly. Visit us at www.prospect-magazine.co.uk. Reprinted with permission.

"A Meaningful Life" by Peter Singer. From *Ethics into Action: Henry Spira and the Animal Rights Movement* (Lanham, Md.: Rowman & Littlefield, 1998). Reprinted with permission.

"Animal Liberation: A Personal View" from *Between the Species,* vol. 2 (Summer 1986), pp. 148–154. Reprinted with permission.

"On Being Silenced in Germany" by Peter Singer. Copyright © 1991 by Peter Singer. Reprinted by permission of the author.

Portions of "An Interview" with Bob Abernethy are from *Religion and Ethics Newsweekly,* September 10, 1999. Reprinted courtesy Thirteen/WNET New York.

Portions of "An Interview" with Nell Boyce are from *New Scientist,* January 8, 2000. Reprinted by permission of *New Scientist,* www.newscientist.com.

Index